CONSTRUCTIONARY®

2nd edition

English-Spanish Construction Dictionary

ALBERTO HERRERA

Constructionary, Second Edition

2 3 4 5 6 7 8 9 TRA/TRA 0 1 2 1 0 9 8 7

ISBN-13: 978-1-58001-378-9
ISBN-10: 1-58001-378-3

COPYRIGHT © 2006, 2000
by
INTERNATIONAL CODE COUNCIL, INC.

PRINTED IN CANADA

Preface

The *Constructionary*™ has become the tool of preference for increasing communication between the Spanish- and English-speaking communities of the construction industry.

The first edition was developed as the result of extensive research done by a team of volunteers and ICC staff members. The success of the *Constructionary* is reflected in its accceptance. Today, it is widely used to improve efficiency and communication on the jobsite. One of the main goals of the *Constructionary* was to unify the use of construction terms in Latin America and the Hispanic community of the U.S. As other dictionaries based on or inspired by this work appear on the market, the *Constructionary* proves its relevance and leads the effort to improve safety, efficiency, quality and uniformity in the construction industry.

This second edition features an extended vocabulary, including new sections of plumbing, electrical and mechanical terms. As in the first edition, useful tables and phrases complement the contents of this practical guide. Contractors, construction workers, engineers, architects, building officials and inspectors will continue to find the *Constructionary* handy in their everyday tasks.

Because language is dynamic, the *Constructionary* is a living document and will continue to grow and improve. Please send your comments and contributions, including any terms you would like to see in future editions of the *Constructionary*, to aherrera@iccsafe.org.

Alberto Herrera
Editor

Acknowledgments

Recognition must be given to Terry Eddy, International Code Council's Senior Vice President of Human Resources, for his original idea of developing a small, practical dictionary of construction terms for the bilingual and the not so bilingual members of the building industry, and to Maria Aragon, ICC's Marketing Manager, for her creativity in naming this bilingual guide.

Our thanks to Benjamin Rodriguez, David Bautista, David Jamieson, Mark Stevens, Miguel Lamas, Philip Ramos, Sergio Barrueto, and Suzane Nunes for their invaluable contributions to the first edition.

We also recognize Paula Chaves, Alberto Iezzi, Raúl Zaradnik and Alejandro Santamaría for their contributions to the plumbing and mechanical sections of this second edition

A special thanks to Mark A. Johnson, ICC's Senior Vice President of Business and Product Development for believing in this work and making possible its development and success.

We recognize the work for the cover done by Duane Acoba as graphic designer and Dianna Hallmark as art director.

Table of Contents

Preface - iii

Acknowledgments - v

Pronunciation Guide - - - - - - - - - - - - - - - - - - viii

English-Spanish A—Z - - - - - - - - - - - - - - - - - 1

Electrical - 93

Plumbing - 101

Mechanical - 113

Tools - 123

Useful Phrases - 131

The Numbers - 139

The Months of the Year - - - - - - - - - - - - - - - - 141

The Days of the Week - - - - - - - - - - - - - - - - - 142

Unit Conversion Tables - - - - - - - - - - - - - - - - 143

Conversion Factors - - - - - - - - - - - - - - - - - - 145

Simplified Pronunciation

This pronunciation guide was developed from a basic communication perspective and not from the international phonetic system.

Accents are indicated by uppercase syllables.

Vowels (5)

A- pronounced   *-AH* as in *father*.

E-   *-EH* as in *mess*

I-   *-EE* as in *see*

O-   *-OH* as the first sound in *owe*

U-   *-OO* as in *boot*

Diphthongs:

The most common used here are:

io-   *-EEOH*

ie-   *-YEH* (the y sounds like *ee)*

ui-   *-WEE*

ua-   *-WAH*

ue-   *-WEH*

Consonants:

Same as in English except:

D is voiced

T and *P* are soft

ny as in *onion*

R soft, <u>*RR*</u> rolled

CH always as in *church*

Abate	**Remover** *(reh-moh-VEHR)* **Anular** *(ah-noo-LAHR)*
Abatement	**Remoción** *(reh-moh-SYOHN)* **Anulación** *(ah-noo-lah-SYOHN)*
Access	**Acceso** *(ahk-SEH-soh)*
Access cover	**Tapa de acceso** *(TAH-pa deh ahk-SEH-soh)* **Cubierta de acceso** *(koo-BYEHR-tah deh ahk-SEH-soh)*
Acoustical tile	**Panel acústico** *(pah-NEL ah-COOS-tee-koh)*
Adapter fitting	**Dispositivo adaptador** *(dees-poh-see-TEE-voh ah-dahp-ta-DOHR)*
Addition	**Ampliación** *(ahm-plee-ah-SYOHN)* **Expansión** *(ehk-spahn-SYOHN)*
Additives and admixtures	**Aditivos y mezclas** *(ah-dee-TEE-vos ee mehs-klahs)*
Air compressor	**Compresor de aire** *(kohm-preh-SOHR deh AH-ee-reh)*
Aisle	**Pasillo** *(pah-SEE-joh)* **Hilera** *(ee-LEH-rah)*
Alley	**Callejón** *(kah-jeh-HOHN)*

Alter	**Modificar** *(moh-dee-fee-KAHR)*
Alteration	**Modificación** *(moh-dee-fee-kah-SYOHN)*
Anchor	**Anclaje** *(ahn-KLAH-heh)*
Anchor bolts	**Pernos de anclaje** *(PEHR-nohs deh ahn-KLAH-heh)* **Tornillos de anclaje** *(tohr-NEE-johs deh ahn-KLAH-heh)*
Angle Iron	**Hierro angular** *(YEH-rroh ahn-goo-LAR)* **Cantonera** *(kahn-toh-NEH-rah)*
Annular grooved nail	**Clavo anular** *(KLAH-voh ah-noo-LAR)*
Antisiphon	**Antisifonaje** *(ahn-tee-see-pho-NAH-heh)*
Apartment house	**Edificio de departamentos** *(eh-dee-FEE-syoh deh deh-par-tah-MEHN-tohs)* **Apartamento residencial** *(ah-par-tah-MEHN-toh reh-see-den-SYAL)*
Approved	**Aprobado** *(ah-proh-BAH-doh)*
Apron	**Repisa** *(reh-PEE-sah)* **Delantal** *(deh-lahn-TAHL)* **Mandil** *(mahn-DEEL)*
Architect	**Arquitecto** *(ahr-kee-tehk-toh)*
Area drain	**Desagüe de área** *(deh-SAH-gweh deh AH-reh-ah)* **Desagüe de patio** *(deh-SAH-gweh deh PAH-lee-oh)* **Resumidero** *(reh-soo-me-DEH-roh)*

Asbestos cement shingle	**Teja de cemento de asbestos** *(TEH-ha deh seh-MEHN-toh deh ahs-BES-tohs)*
	Tablilla de fibrocemento *(tah-BLEE-jah deh fee-broh-seh-MEHN-toh)*
Asphalt	**Asfalto** *(ahs-FAHL-toh)*
Atrium	**Atrio** *(AH-tree-oh)*
Attic	**Tapanco** *(tah-PAHN-koh)*
	Entrepiso *(ehn-treh-PEE-soh)*
Automatic	**Automático** *(ah-oo-toh-MAH-tee-koh)*
Automatic closing device	**Dispositivo de cierre automático** *(dees-poh-see-TEE-voh deh SYEH-rreh ah-oo-toh-MAH-tee-koh)*
Automatic fire sprinkler system	**Sistema de rociadores automáticos** *(sees-TEH-mah deh roh-syah-DOH-rehs ah-oo-toh-MAH-tee-kohs)*
Automatic fire-extinguishing system	**Sistema automático de extinción de incendios** *(sees-TEH-mah ah-oo-to-MAH-tee-koh deh ehk-steen-SYOHN deh een-SEHN-dee-ohs)*
Awnings	**Toldos** *(TOHL-dohs)*

3

Back hoe

Retroexcavadora
(reh-troh-ex-kah-vah-DOH-rah)

Excavadora
(ex-kah-vah-DOH-rah)

Backfill

Relleno
(reh-JEH-noh)

Backflow

Contraflujo
(kohn-trah-FLOO-hoh)

Backflow preventer

Válvula de contraflujo
(VAHL-voo-lah deh kohn-trah-FLOO-hoh)

Backing

Soporte
(soh-POHR-teh)

Respaldo
(rehs-PAHL-doh)

Balcony

Balcón
(bahl-KOHN)

Ball cock

Válvula de flotador
(VAHL-voo-lah deh floh-tah-DOHR)

Flotador
(floh-tah-DOHR)

Llave de flotador
(JAH-veh deh floh-tah-DOHR)

Ball valve

Llave de flujo
(JAH-veh deh FLOO-hoh)

Band joist

Viga lateral
(VEE-gah lah-teh-RAHL)

Bar

Barra
(BAH-_rr_ah)

Barreta
(bah-_RR_EH-tah)

Barbed nail

Clavo afilado
(KLAH-voh ah-fee-LAH-doh)

Baseboard	**Zócalo** *(SOH-kah-loh)*
Basement	**Sótano** *(SOH-tah-noh)*
	Subterráneo *(soob-teh-<u>RR</u>AH-neh-oh)*
Bathroom	**Cuarto de baño** *(KWAR-to deh BAH-nyoh)*
	Sanitario *(sah-nee-TAH-ryoh)*
Bathroom sink	**Lavabo** *(lah-VAH-boh)*
Bathtub	**Bañera** *(bah-NYEH-rah)*
	Tina de baño *(TEE-nah deh BAH-nyoh)*
	Bañadera *(bah-NYAH-deh-rah)*
Batten	**Rastrillo** *(rahs-TREE-joh)*
	Cubrejuntas *(koo-breh-HOON-tahs)*
	Listón travesaño *(lees-TON trah-veh-SAH-nyoh)*
Battery	**Batería** *(bah-teh-REE-ah)*
	Pila *(PEE-lah)*
Beam	**Viga** *(VEE-gah)*
Bearing load	**Carga de cojinete** *(KAHR-gah deh koh-hee-NEH-teh)*
Bedroom	**Cuarto** *(KWAR-toh)*
	Habitación *(ah-bee-tah-SYOHN)*
	Recámara *(re-KAH-mah-rah)*
	Dormitorio *(dor-mee-TOH-ree-oh)*
Beech	**Haya** *(AH-jah)*

5

Below-grade walls	**Muros por debajo del nivel de terreno** *(MOO-rohs pohr deh-BAH-ho dehl nee-VEHL dehl teh-RREH-noh*
Bird's mouth	**Incisión recta** *(een-see-SYOHN REHK-ta)*
	Muesca *(MWESS-kah)*
	Ranura *(rah-NOO-rah)*
	Rebajo *(reh-BAH-hoh)*
Bleachers	**Tribunas** *(tree-BOON-ahs)*
	Gradas *(GRAH-dahs)*
Blind nailed	**Con clavos ocultos** *(kohn KLAH-vohs oh-COOL-tohs)*
Block, Blocking	**Trabas** *(TRAH-bahs)*
	Trabar *(trah-BAHR)*
	Bloque *(BLOH-keh)*
	Bloquear *(bloh-KEAHR)*
Board	**Panel** *(PAH-nehl or pah-NEHL)*
	Tabla *(TAH-blah)*
	Tablero *(TAH-bleh-roh)*
Boiler	**Caldera** *(kahl-DEH-rah)*
	Calentador *(kah-len-tah-DOHR)*
Boiler room	**Cuarto de calderas** *(KWAR-toh deh kahl-DEH-rahs)*
Bolt	**Perno** *(PEHR-noh)*
	Tornillo *(tohr-NEE-joh)*

6

Bonding jumper	**Borne de enlace** *(BOHR-neh deh ehn-LAH-seh)*
	Terminal de enlace *(tehr-mee-NAHL deh ehn-LAH-seh)*
Box nail	**Clavo para madera** *(KLAH-voh PAH-rah mah-DEH-rah)*
	Clavo de cabeza grande plana *(KLAH-voh deh kah-BEH-sah GRAHN-deh PLAH-nah)*
Brace	**Tirante** *(tee-RAN-teh)*
Bracing	**Arriostramiento** *(ah-rryohs-trah-MYEN-toh)*
Braced frame	**Estructura arriostrada** *(ehs-trook-TOO-rah ah-rryohs-TRAH-dah)*
	Pórtico arriostrado *(POHR-tee-koh ah-rryohs-TRAH-doh)*
Bracket	**Brazo** *(BRAH-soh)*
	Ménsula *(MEHN-soo-lah)*
Branch	**Ramal** *(rah-MAHL)*
Brass	**Bronce** *(BROHN-seh)*
	Latón *(lah-TOHN)*
Braze	**Soldar en fuerte** *(sohl-DAR ehn FWER-teh)*
Brazing alloy	**Aleación para soldar** *(ah-leh-ah-SYOHN PAH-rah sol-DAR)*
Brazing flux	**Fundente para soldar** *(foon-DEN-teh PAH-rah sol-DAR)*
Brick	**Ladrillo** *(lah-DREE-joh)*
Bridging	**Puntales de refuerzo** *(poon-TAH-les deh reh-FWEHR-soh)*
Brown coat	**Revoque** *(reh-VOH-keh)*

7

Building   **Edificación**
                                (eh-dee-fee-kah-SYOHN)

                                Edificio
                                (eh-dee-FEE-syoh)

Building department   **Departamento de**
                                construcción/edificación
                                (deh-par-tah-MEN-toh deh kohn-strook-SYOHN/eh-dee-fee-kah-SYOHN)

                                Departamento de obras de
                                edificación
                                (deh-par-tah-MEN-toh deh OH-brahs deeh-dee-fee-kah-SYOHN)

                                Departamento de obras públi-cas/privadas
                                (deh-par-tah-MEN-toh deh OH-brahs POO-blee-kahs/pree-VAH-dahs)

Building drain   **Desagüe de la edificación/**
                                del edificio
                                (deh-SAH-gweh deh lah eh-dee-fee-kah-SYOHN/dehl eh-dee-FEE-syoh)

                                Resumidero
                                (reh-soo-me-DEH-roh)

Building documentation   **Documentación de obra/**
                                ingeniería
                                (doh-koo-mehn-tah-SYOHN deh OH-brah/een-he-nyeh-REE-ah)

Building inspector   **Inspector de obras**
                                (eens-pek-TOHR deh OH-brahs)

                                Inspector de construcción
                                (eens-pek-TOHR deh kohn-strook-SYOHN)

Building official   **Director de obras**
                                (dee-rek-TOHR deh OH-brahs)

                                Jefe de obras
                                (HEH-feh deh OH-brahs)

                                Autoridad competente
                                (ah-oo-toh-ree-DAHD kohm-peh-TEHN-teh)

Building site   **Terreno de obra**
                                (teh-RREH-noh deh OH-brah)

                                Sitio de construcción
                                (SEE-tee-oh deh kohn-strook-SYOHN)

                                Obra de construcción
                                (OH-brah deh kohn-strook-SYOHN)

Built-up roofing   **Cubierta de techo compuesta**
                                (koo-BYEHR-tah deh TEH-choh kohm-PWFHS-tah)

Bundle

Atado
(ah-TAH-doh)

Bulto
(BOOL-toh)

Bushing

Manguito
(mahn-GUEE-toh)

Forro
(BOOL-toh)

Butt joint

Junta a tope
(HOON-tah ah TOH-peh)

Junta de cubrejunta
(HOON-tah deh koo-breh-HOON-tah)

Bypass

Derivación
(deh-ree-vah-SYOHN)

Desvío
(dehs-VEE-oh)

Paso
(PAH-soh)

Cabinet
Gabinete
(gah-bee-NEH-teh)

Cabinetmaker
Ebanista
(eh-bah-NEES-tah)

Cable tray
Bandeja portacables
(bahn-DEH-hah pohr-tah-KAH-blehs)

Caissons
Cajones de aire comprimido
(kah-HOH-nehs deh ah-ee-reh kohm-pree-MEE-doh)

Canopy
Toldo
(TOHL-doh)

Cubierta
(koo-BYER-tah)

Cantilever
Voladizo
(volah-DEE-soh)

Carpenter
Carpintero
(kahr-peen-TEH-roh)

Cartridge fuse
Fusible de cartucho
(foo-SEE-bleh deh kahr-TOO-choh)

Casing nail
Clavo de cabeza perdida
(KLAH-voh deh kah-BEH-sah pehr-DEE-dah)

Cast stone
Piedra moldeada
(PYEH-drah mohl-deh-AH-dah)

Piedra de sillar
(PYEH-drah deh see-JAHR)

Caulking
Masillado
(mah-se-JAH-doh)

Masillar
(mah-see-JAHR)

Cavity wall
Muro hueco
(moo-roh WEH-koh)

Ceiling	**Cielorraso** *(SYEH-loh-RRAH-soh)*
Ceramic floor	**Piso cerámico** *(PEE-soh seh-RAH-mee-koh)*
Ceramic tile	**Baldosas cerámicas** *(bahl-DOH-sahs seh-RAH-mee-kahs)*
Certificate of occupancy	**Certificado de uso** *(sehr-tee-fee-KAH-doh deh OO-soh)*
Chain	**Cadena** *(kah-deh-nah)*
Chalk line	**Linea de marcar** *(LEE-neh-ah deh mahr-KAHR)* **Tendel** *(tehn-DEHL)* **Linea de gis** *(LEE-neh-ah de HEES)*
Chase	**Canaletas** *(kah-nah-LEH-tahs)* **Muesca** *(moo-EHS-kah)*
Check valve	**Válvula de contraflujo** *(VAHL-voo-lah deh kohn-trah-FLOO-hoh)*
Chimney	**Chimenea** *(chee-meh-NEH-ah)*
Chimney chase	**Acanaladura de chimenea** *(ah-kah-nah-lah-DOO-rah deh chee-meh-NEH-ah)*
Chimney, factory-built	**Chimenea prefabricada** *(chee-meh-NEH-ah preh-fah-bree-KAH-dah)*
Chimney liner	**Revestimiento de chimenea** *(reh-vehs-tee-MYEHN-toh deh chee-meh-NEH-ah)*
Chute	**Ducto** *(DOOK-toh)*
Chute, linen	**Ducto de lencería** *(DOOK-toh deh lehn-seh-REE-ah)*
Chute, rubbish	**Ducto de basura** *(DOOK-toh deh bah-soo-rah)*

Cinder	**Cenizas** *(seh-NEE-sahs)*
Circuit	**Circuito** *(seer-KWEE-toh)*
Circuit breaker	**Apagador** *(ah-pah-gah-DOR)* **Interruptor automático** *(een-teh-rroop-TOHR ah-oo-toh-MAH-tee-koh)* **Interruptor de circuito** *(een-teh-rroop-TOHR de seer-KWEE-toh)*
Circuit breaker panel	**Cuadro de cortacircuito** *(KWAH-droh dehl kor-tah-seer-KWEE-toh)* **Tablero de cortacircuito** *(tah-BLEH-roh deh kor-tah-seer-KWEE-toh)*
Cistern	**Aljibe** *(ahl-HE-beh)* **Cisterna** *(sees-TEHR-nah)*
Clay	**Arcilla** *(ar-SEE-jah)* **Barro** *(BAH-rroh)*
Cleanout	**Registro** *(reh-HEES-troh)* **Abertura de limpieza** *(ah-behr-TOO-rah deh leem-PYEH-zah)*
Clearance	**Espacio libre** *(ehs-PAH-syoh LEE-bre)*
Clear span	**Luz libre** *(loos LEE-breh)* **Claro** *(CLAH-roh)*
Coal	**Hulla** *(OO-jah)* **Carbón** *(kahr-BOHN)*
Coarse	**Grueso** *(GRWEH-soh)* **Áspero** *(AHS-peh-roh)*

Code	**Código** *(KOH-dee-goh)*
Code official	**Autoridad competente** *(ah-oo-toh-ree-DAHD kohm-peh-TEHN-teh)* **Director/Jefe de obras** *(dee-rek-TOHR/HEH-feh deh OH-brahs)* **Oficial de códigos** *(oh-fee-SYAHL deh KOH-dee-gohs)*
Collar joint	**Junta de collar** *(HOON-tah deh koh-JAHR)*
Column	**Columna** *(koh-LOOM-nah)*
Combination fixture	**Artefacto de combinación** *(ahr-teh-FAHK-toh deh kohm-bee-nah-SYOHN)* **Mueble de combinación** *(MWEH-bleh deh kohm-bee-nah-SYOHN)*
Combustible liquid	**Líquido combustible** *(LEE-kee-doh kohm-boos-TEE-bleh)*
Compression coupling	**Acoplamiento de compresión** *(ah-koh-plah-MYEHN-toh deh kohm-preh-SYOHN)* **Empalme de compresión** *(ehm-PAHL-meh deh kohm-preh-SYOHN)* **Manguito** *(mahn-GY-toh)* **Manchón de manguito** *(mahn-CHON deh mahn-GY-toh)*
Concealed spaces	**Espacios ocultos** *(ehs-PAH-syohs oh-KOOL-tohs)* **Cavidades ocultas** *(kah-vee-DAH-dehs oh-KOOL-tahs)*
Concrete	**Hormigón** *(ohr-mee-GOHN)* **Concreto** *(kohn-KREH-toh)*
Concrete cover	**Cubierta de hormigón/concreto** *(koo-BYEHR-tah deh ohr-mee-GOHN/kohn-KREH-toh)*
Condominium	**Condominio residencial** *(kohn-doh-MEE-nee-oh reh-see-den-SYAL)*

13

Conductor	**Conductor** *(kohn-dook-TOHR)*
Conductor wire	**Alambre conductor** *(ah-LAHM-breh kohn-dook-TOHR)*
Conduit	**Conducto** *(kohn-DOOK-toh)*
	Tubería *(too-beh-REE-ah)*
Congregate residence	**Residencia comunitaria** *(reh-see-DEN-syah koh-moo-nee-TAH-ree-ah)*
Connection	**Conexión** *(koh-nek-SYOHN)*
	Unión *(oon-YOHN)*
Connector	**Conector** *(koh-nek-TOHR)*
Construction Health and Safety	**Higiene y seguridad en la construcción** *(ee-hee-EH-neh ee seh goo-ree-DAHD ehn lah kohn-strook-SYOHN)*
Construction joint	**Junta de construcción** *(HOON-tah deh kohn-strook-SYOHN)*
Construction schedule (CPM)	**Cronograma de construcción** *(kroh-noh-GRAH-mah deh kohn-strook-SYOHN)*
	Plan de avance de obra *(PLAHN deh ah-VAHN-seh deh OH-brah)*
Contamination	**Contaminación** *(kohn-tah-mee-nah-SYOHN)*
Contraction joint	**Junta de contracción** *(HOON-tah deh kohn-trac-SYOHN)*
Contractor	**Constructor** *(kohn-strook-TOHR)*
	Contratista *(kohn-trah-TEE-stah)*
Coping	**Albardilla** *(ahl-bar-DEE-jah)*

Mojinete
(mo-hee-NEH-teh)

Cumbrera
(coom-BREH-rah)

Caballete
(kah-bah-JEH-teh)

Copper, hard drawn   **Cobre estirado en frío**
(KOH-breh ehs-tee-RAH-doh ehn FREE-oh)

Copper, sheet   **Lámina de cobre**
(LAH-mee-nah deh KOH-breh)

Copper, wrought   **Cobre forjado**
(KOH-breh fohr-HAH-doh)

Cornerite   **Guardaesquinas**
(gwar-dah-ehs-KEE-nahs)

Cornices   **Cornisas**
(kor-NEE-sahs)

Corrosion-resistant   **Anticorrosivo**
(ahn-tee-koh-rroh-SEE-voh)

Resistente a la corrosión
(reh-sees-TEHN-teh ah lah koh-rroh-SYOHN)

Corrosive   **Corrosivo**
(koh-rroh-SEE-voh)

Coupling   **Acoplamiento**
(ah-koh-plah-MYEHN-toh)

Copla
(KOH-plah)

Manguito
(man-GY-toh)

Coupling beams   **Vigas de acoplamiento**
(VEE-gahs deh ah-koh-plah-MYEHN-toh)

Court   **Patio interno**
(PAH-tee-oh een-TEHR-noh)

Cover, Covering   **Recubrimiento**
(reh-koo-bree-MYEHN-toh)

Tapadera
(tah-pah-DEH-rah)

Revestimiento
(reh-veh-stee-MYEHN-toh)

Cubierta
(coo-BYEHR-tah)

15

Cracked walls	**Muros rajados** *(moo-rohs rah-HA-dohs)* **Paredes rajadas** *(pah-REH-dehs rah-HA-dahs)*
Cracks	**Rajadas** *(rah-HA-dahs)* **Grietas** *(GRYEH-tahs)* **Partidas** *(pahr-TEE-dahs)*
Crawl space	**Espacio angosto** *(ehs-PAH-syoh ahn-GOHS-toh)* **Sótano de poca altura** *(SOH-tah-noh deh POH-kah ahl-TOO-rah)*
Cripple wall	**Muro corto** *(MOO-roh KOHR-toh)*
Cross connection	**Conexión cruzada** *(koh-nek-SYOHN kroo-SAH-dah)*
Cross-grain	**Fibra transversal** *(FEE-brah trans-vehr-SAL)* **Contragrano** *(kohn-trah-GRAH-noh)*
Crown	**Corona (grapas)** *[koh-ROH-nah (GRAH-pahs)]*
Crown face	**Cara combada** *(KAH-rah com-BAH-dah)*
Culvert	**Alcantarilla** *(alkan-tah-REE-jah)* **Desagüe** *(deh-SAH-gweh)*
Curb	**Guarnición** *(gwahr-nee-SYOHN)* **Cordón** *(kohr-DON)* **Flanco** *(FLAHN-koh)* **Bordillo** *(bohr-DEE-joh)*
Cut-off valve	**Válvula de cierre** *(VAHL-voo-lah deh SYEH-rreh)*

Damper	**Regulador**
	(reh-goo-lah-DOHR)
Dead end	**Terminal**
	(tehr-mee-NAHL)
	Extremos cerrados
	(ehk-STREH-mohs seh-RRAH-dohs)
	Sin salida
	(seen sah-LEE-dah)
Dead load	**Carga muerta**
	(KAHR-gah MWEHR-tah)
	Carga permanente
	(KAHR-gah pehr-mah-NEHN-teh)
Deck, Decking	**Cubierta**
	(koo-BYEHR-tah)
Design drawings	**Planos**
	(PLAH-nohs)
Designer	**Diseñador**
	(dee-seh-nyah-DOHR)
Detached building	**Edificación separada**
	(eh-dee-fee-kah-SYOHN seh-pah-RAH-dah)
Device	**Dispositivo**
	(dees-poh-see-TEE-voh)
Diagonal bracing	**Arriostramiento diagonal**
	(ah-rryohs-trah-MYEHN-toh dyah-goh-NAHL)
Dig	**Excavar**
	(ex-kah-VAHR)
Dilapidation	**Deterioro**
	(deh-teh-RYOH-roh)
Disability	**Discapacidad**
	(dees-kah-pah-see-DAHD)

Discharge pipe   **Tubo de descarga**
                              (TOO-boh deh dehs-KAHR-gah)

Discontinuous beams   **Vigas discontinuas**
                              (VEE-gahs dees-kohn-TEE-nwahs)

Dispense   **Trasvasar**
                              (trahs-vah-SAHR)

Dispersal area, safe   **Área segura de dispersión**
                              (AH-reh-ah seh-GOO-rah deh dees-pehr-SYOHN)

Displacement   **Corrimiento**
                              (koh-rree-MYEHN-toh)

                              Desplazamiento
                              (dehs-plah-sah-MYEHN-toh)

Door   **Puerta**
                              (PWER-tah)

Door assemblies   **Sistemas de puertas**
                              (sees-TEH-mahs deh PWEHR-tahs)

Door sill   **Umbral**
                              (oom-BRAL)

Doorbell   **Timbre**
                              (TEEM-breh)

Doorway   **Claro de puerta**
                              (KLAH-roh deh PWEHR-tah)

                              Entrada
                              (ehn-TRAH-dah)

                              Portal
                              (pohr-TAHL)

Dormer   **Buharda**
                              (boo-AHR-dah)

                              Buhardilla
                              (boo-ahr-DEE-jah)

Dormitory   **Residencias para estudiantes**
                              (reh-see-DEN-syahs PAH-rah ehs-too-DYAHN-tehs)

                              Dormitorio estudiantil
                              (dor-mee-TOH-ree-oh ehs-too-dyahn-TEEL)

Double plate   **Solera doble**
                              (soh-LEH-rah DOH-bleh)

Double pole breaker   **Interruptor automático bipolar**
                              (een-teh-rroop-TOHR ah-oo-toh-MAH-tee-koh bee-poh-LAHR)

Doubled	**Adosado** *(ah-doh-SAH-doh)*
Doubler plates	**Placas** *(PLAH-kahs)*
	Placas de refuerzo *(PLAH-kahs deh reh-FWEHR-soh)*
Draft stop	**Cierre de tiro** *(SYEH-rreh deh TEE-roh)*
	Barrera contra corriente de aire *(ba-RREH-rah kohn-trah koh-RRYEHN-teh deh ah-ee-reh)*
Draftsman	**Dibujante** *(dee-boo-HAHN-teh)*
Drain, Drainage	**Desagüe** *(deh-SAH-gweh)*
	Drenaje *(dreh-NAH-heh)*
Drawn	**Extruído** *(eks-troo-EE-doh)*
Dressing room	**Vestidor** *(vehs-tee-DOHR)*
Drilling	**Perforación** *(pehr-foh-rah-SYOHN)*
Driven	**Impulsado** *(eem-pool-SAH-doh)*
Dry wall	**Muro en seco** *(MOO-roh ehn SEH-koh)*
	Tablero de yeso *(tah-bleh-roh deh YEH-soh)*
Dryer	**Secadora** *(seh-kah-DOH-rah)*
Dual system	**Sistema doble** *(sees-TEH-mah DOH-bleh)*
Duct	**Conducto** *(kohn-DOOK-toh)*
Dumbwaiter	**Montacargas** *(mohn-tah-KAHR-gahs)*
	Montaplatos *(mohn-tah-PLAH-tohs)*

Dwelling

Vivienda
(vee-VYEHN-dah)

Residencia
(reh-see-DEN-syah)

Habitación
(ah-bee-tah-SYOHN)

Dwelling unit

Unidad de vivienda
(oo-nee-DAHD deh vee-VYEHN-dah)

Unidad habitacional
(oo-nee-DAHD ah-bee-tah-syo-NAL)

Earth work	**Terraplén** *(teh-rrah-PLEN)*
Earthquake load	**Carga sísmica** *(KAHR-gah SEES-mee-kah)*
Eave	**Alero** *(ah-LEH-roh)*
Edge (on edge)	**Canto (de canto a canto)** *[KAHN-toh (deh KAHN-toh ah KAHN-toh)]* **Borde** *(BOHR-deh)*
Egress	**Salida** *(sah-LEE-dah)*
Electrical fixture	**Artefactos eléctricos** *(ahr-teh-FAHK-toh eh-LEHK-tree-kohs)*
Electrical outlet	**Enchufe** *(ehn-CHOO-feh)* **Tomacorriente** *(Toh-mah-koh-RRYEHN-teh)*
Electrician	**Electricista** *(eh-lehk-tree-SEES-tah)*
Electricity	**Electricidad** *(eh-lehk-tree-see-DAHD)*
Elevator	**Ascensor** *(ah-sehn-SOHR)* **Elevador** *(eh-leh-vah-DOHR)*
Elevator car	**Coche de ascensor** *(KOH-cheh deh ah-sehn-SOHR)*
Embankment	**Terraplén** *(teh-rrah-PLEN)*
Embedded	**Empotrados** *(ehm-poh-TRAH-dohs)*

	Incrustado *(een-kroos-TAH-doh)*
	Arraigado *(ah-<u>rr</u>ah-ee-GAH-doh)*
Embedment	**Empotradura** *(ehm-poh-trah-DOO-rah)*
Enclose	**Encerrar** *(ehn-seh-<u>RR</u>AHR)*
Enclosed	**Encerrado** *(ehn-seh-<u>RR</u>AH-doh)*
Enclosure	**Cerramiento** *(seh-<u>rr</u>ah-MYEHN-toh)*
Encompass	**Incluir** *(een-kloo-EER)*
Ends	**Puntas** *(POON-tahs)*
	Extremos *(ehks-TREH-mohs)*
Enforce	**Hacer cumplir** *(ah-SEHR koom-PLEER)*
Engineer	**Ingeniero** *(een-heh-NYEH-roh)*
Escalator	**Escalera mecánica** *(ehs-kah-LEH-rah meh-KAH-nee-kah)*
Essential facilities	**Instalaciones esenciales** *(eens-tah-lah-SYOH-nehs eh-sehn-SYAH-lehs)*
	Edificaciones esenciales *(eh-dee-fee-kah-SYOH-nehs eh-sehn-SYAH-lehs)*
Exhaust	**Escape** *(ehs-KAH-peh)*
	Extracción *(ehks-trahk-SYOHN)*
Exhaust fan	**Ventilador de extracción** *(vehn-tee-lah-DOHR deh ehks-trahk-SYOHN*
Exit	**Salida** *(sah-LEE-dah)*

Exit door

Puerta de salida
(PWEHR-tah deh sah-LEE-dah)

Expansion bolt

Perno de expansión
(PEHR-noh deh ehk-spahn-SYOHN)

Tornillo de expansión
(tohr-NEE-joh deh ehk-spahn-SYOHN)

Expansion joint

Junta de dilatación/expansión
(HOON-tah deh dee-lah-tah-SYOHN/ehk-spahn-SYOHN)

Extension cord

Cable de extensión
(KAH-bleh deh ehk-stehn-SYOHN)

Exterior wall

Muro/Pared exterior
(MOO-roh/pa-REHD ehk-steh-RYOHR)

Exterior/interior surface

Superficie exterior/interior
(soo-per-FEE-syeh ehk-steh-RYOHR/een-teh-RYOHR)

Façade	**Alzado** *(al-ZAH-doh)* **Fachada** *(fah-CHAH-dah)*
Face grain	**Veta superficial** *(VEH-tah soo-pehr-fee-SYAHL)*
Face load	**Carga ditribuida** *(KAHR-gah dees-tree-boo-EE-dah)*
Face mount hangers	**Percha de montaje frontal** *(PEHR-chah deh mon-tah-HEH fron-TAHL)*
Face shield	**Careta** *(kah-REH-tah)*
Facing brick	**Ladrillos para frentes** *(lah-DREE-johs PAH-rah FREHN-tehs)*
Factory built	**Prefabricado** *(preh-fah-bree-KAH-doh)*
Factory-made	**Hecho en fábrica** *(EH-choh ehn FAH-bree-kah)*
Fan	**Ventilador** *(vehn-tee-lah-DOHR)* **Abanico** *(ah-bah-NEE-koh)*
Fasteners	**Anclajes** *(ahn-KLAH-hehs)*
Faucet	**Llave** *(JAH-veh)* **Grifo** *(GREE-foh)*
Feeder cable	**Cable de alimentación** *(KAH-bleh de ah-lee-mehn-tah-SYOHN)*
Felt	**Fieltro** *(FYEHL-troh)*

Fence

Cerca
(SEHR-ka)

Barda
(BAR-dah)

Mediera
(meh-DYEH-rah)

Fiberboard

Tablero de fibra
(tah-BLEH-roh deh FEE-brah)

Field test

Prueba en obra
(PRWEH-bah ehn OH-brah)

Filled

Rellenado
(reh-jeh-NAH-doh)

Finish

Acabado
(ah-kah-BAH-doh)

Terminado
(tehr-mee-NAH-doh)

Finishing nail

Clavo sin cabeza
(KLAH-voh seen kah-BEH-sah)

Fire alarm system

Sistema de alarma contra incendios
(sees-TEH-mah deh ah-LAR-mah kohn-trah een-SEHN-dee-ohs)

Fire Code

Código de Incendios
(KOH-dee-goh deh een-SEHN-dee-ohs)

Fire department

Cuerpo/Departamento de bomberos
(KWER-poh/deh-par-tah-MEHN-toh deh bohm-BEH-rohs)

Fire department access

Acceso para bomberos
(ahk-SEH-soh PAH-rah bohm-BEH-rohs)

Fire department connection

Toma de impulsión
(TOH-mah deh eem-pool-SYOHN)

Conexión para bomberos
(koh-nek-SYOHN PAH-ra bohm-BEH-rohs)

Fire extinguisher

Extinguidor
(ehk-steen-ghee-DOHR)

Extintor
(ehk-steen-TOHR)

Matafuegos
(mah-tah-FWEH-gohs)

Fireblock	**Bloque antifuego** *(BLOH-keh ahn-tee-FWEH-goh)*
Fireblocking	**Bloqueado antifuego** *(bloh-keh-AH-do ahn-tee-FWEH-goh)*
Firebox	**Fogón** *(foh-GOHN)*
Firebrick	**Ladrillo de fuego** *(lah-DREE-joh deh FWEH-goh)*
Fireplace	**Chimenea** *(chee-meh-NEH-ah)*
	Hogar *(oh-GAHR)*
First floor/story	**Primer piso** *(pree-MER PEE-soh)*
Fish tape	**Cinta pescadora** *(SEEN-tah pehs-kah-DOH-rah)*
Fitting	**Accesorio** *(ahk-seh-SOH-ryoh)*
	Conexión *(koh-nek-SYOHN)*
Fixture	**Artefacto** *(ahr-teh-FAHK-toh)*
	Accesorio *(ahk-seh-SOH-ryoh)*
Fixture, bathroom	**Artefacto sanitario** *(ahr-teh-FAHK-toh sah-nee-TAH-ryoh)*
Fixture trap	**Trampa hidráulica** *(TRAM-pah ee-DRAH-oo-lee-kah)*
	Sifón *(see-FOHN)*
	Trampa de artefacto *(TRAM-pah deh ahr-teh-FAHK-toh)*
Flammable	**Inflamable** *(een-flah-MAH-bleh)*
Flammable liquid	**Líquido inflamable** *(LEE-kee-doh een-flah-MAH-bleh)*
Flammability	**Inflamabilidad** *(een-flah-mah-bee-lee-DAHD)*
Flanges	**Patines** *(pah-TEE-nehs)*

Bridas
(BREE-dahs)

Alas
(AH-lahs)

Flashing

Cubrejuntas
(koo-breh-HOON-tahs)

Tapajuntas
(tah-pah-HOON-tahs)

Plancha de escurrimiento
(PLAHN-cha deh ehs-koo-rree-MYEHN-toh)

Vierteaguas
(vee-er-teh-AH-gwahs)

Botaguas
(boh-TAH-gwahs)

Flashlight

Linterna
(leen-TEHR-nah)

Lámpara
(LAHM-pah-rah)

Flex conduit

Conducto portacables flexible
(kohn-DOOK-toh pohr-tah-KAH-blehs flek-SEE-bleh)

Floodlight

Iluminación industrial
(ee-loo-mee-nah-SYOHN een-doos-TRYAHL)

Foco industrial
(FOH-koh een-doos-TRYAHL)

Floor

Piso
(PEE-soh)

Solado
(soh-LAH-doh)

Floor deck

Plataforma
(plah-tah-FOHR-mah)

Floor tile (small slab)

Loseta
(loh-SEH-tah)

Flooring

Revestimientos para pisos
(reh-vehs-tee-MYEHN-tohs PAH-rah PEE-sohs)

Material para pisos
(mah-teh-RYAHL PAH-rah PEE-sohs)

Flue

Conductos de humo
(kohn-DOOK-tohs deh OO-moh)

Fluorescent

Fluorescente
(floo-oh-reh-SEHN-teh)

Footboard	**Tabla de piso** *(TAH-blah deh PEE-soh)*
Footing	**Zapata** *(sah-PAH-tah)*
	Zarpa *(SAR-pah)*
	Cimiento *(see-MYEHN-toh)*
Foreman	**Capataz** *(kah-pah-TAHS)*
	Sobrestante *(soh-breh-STAHN-teh)*
	Supervisor *(soo-pehr-vee-SOHR)*
Forms (concrete)	**Encofrados** *(ehn-koh-FRAH-dohs)*
Formwork	**Encofrado** *(ehn-koh-FRAH-doh)*
Foundation	**Fundación** *(foon-dah-SYOHN)*
	Cimentación *(see-men-tah-SYOHN)*
Foundation sill plate	**Placa de solera de fundación** *(PLAH-kah deh soh-LEH-rah deh foon-dah-SYOHN)*
Foundation walls	**Muros de fundación** *(MOO-rohs deh foon-dah-SYOHN)*
Frame	**Marco** *(MAR-koh)*
	Estructura *(ehs-trook-TOO-rah)*
	Pórtico *(POHR-tee-koh)*
	Bastidor *(bahs-tee-DOHR)*
	Armazón *(ar-mah-SOHN)*
Frame, door/window	**Marco de puerta/ventana** *(mar-koh deh PWEHR-tah/vehn-TAH-nah)*
Framed	**Armado** *(ar-MAH-doh)*

Framework   **Armazón**
                       (ar-mah-SOHN)

Framing   **Estructura**
                       (ehs-trook-TOO-rah)

Fumes   **Gases**
                       (GAH-sehs)

Furred out, Furring **Enrasado**
                       (ehn-rah-SAH-doh)

Fuse   **Fusible**
                       (foo-SEE-bleh)

Fuse box   **Caja de fusibles**
                       (KAH-hah deh foo-SEE-blehs)

G

Gable
Hastial
(ahs-TYAHL)

Gable construction
Construcción a dos aguas
(kohns-trook-SYOHN ah dohs AH-gwahs)

Gable roof
Techo a dos aguas
(TEH-cho ah dos AH-gwahs)

Gable rake
Cornisa inclinada
(kor-NEE-sah een-klee-NAH-dah)

Gage/Gauge
(instrument)
Manómetro
(mah-NOH-meh-troh)

Indicador
(een-dee-kah-DOHR)

Gage/Gauge
(thickness)
Calibre
(kah-LEE-breh)

Espesor
(ehs-peh-SOHR)

Garage
Garaje
(gah-RAH-heh)

Cochera
(koh-CHEH-rah)

Garbage disposal
**Triturador de basura/
desperdicios**
*(tree-too-rah-DOHR deh bah-SOO-rah/
dehs-pehr-DEE-syohs)*

Gasket
Arandela
(ar-ahn-DEH-lah)

Empaque
(ehm-PAH-keh)

Junta
(HOON-tah)

Gas-fired
Alimentado por gas
(ah-lee-men-TAH-do pohr gahs)

Gas main	**Conducto/cañería principal de gas** *(kohn-DOOK-toh/kah-nyeh-REE-ah preen-see-PAL deh GAHS)*
Gate valve	**Llave de paso** *(JAH-veh deh PAH-soh)*
Generator	**Generador** *(heh-neh-rah-DOHR)*
Girder	**Viga maestra** *(VEE-gah mah-EHS-trah)* **Viga principal** *(VEE-gah preen-see-PAL)* **Viga** *(VEE-gah)* **Jácena** *(HAH-seh-nah)*
Glazed, Glazing	**Vidriado** *(vee-DRYAH-doh)* **Encristalado** *(ehn-krees-tah-LAH-doh)*
Glass Brick	**Ladrillo de vidrio** *(lah-DREE-joh deh VEE-dryoh)*
Glue	**Pegamento** *(peh-gah-MEHN-toh)*
Grab bars	**Barras de apoyo** *(BAH-rrahs deh ah-POH-joh)* **Barra de soporte** *(BAH-rrahs deh soh-POHR-teh)* **Agarraderas** *(ah-gah-rrah-DEH-rahs)*
Grade	**Nivel de terreno** *(nee-VEHL deh teh-RREH-noh)* **Rasante** *(rah-SAHN-teh)*
Grade beam	**Viga de fundación** *(VEE-gah deh foon-dah-SYOHN)*
Graded lumber	**Madera elaborada** *(mah-DEH-rah eh-lah-boh-RAH-dah)* **Madera clasificada** *(mah-DEH-rah klah-see-fee-KAH-dah)*
Grading	**Nivelación de terreno** *(nee-veh-lah-SYOHN deh teh-RREH-noh)*

Grandstands	**Tribunas** *(tree-BOON-ahs)*
	Gradas *(GRAH-dahs)*
Gravel	**Grava** *(GRAH-vah)*
	Cascajo *(kahs-KAH-hoh)*
	Ripio *(REE-pee-oh)*
Grease interceptor	**Interceptor de grasas** *(een-tehr-sehp-TOHR deh GRAH-sahs)*
Grease trap	**Colector de grasas** *(koh-lehk-TOHR deh GRAH-sahs)*
Grille	**Rejilla** *(reh-HEE-jah)*
	Reja *(REH-hah)*
Grip	**Agarre** *(ah-GAH-rreh)*
Grommet	**Arandela** *(ar-ahn-DEH-lah)*
	Ojal *(oh-HAHL)*
Groove	**Ranura** *(rah-NOO-rah)*
Gross area	**Área total** *(AH-reh-ah toh-TAHL)*
	Área bruta *(AH-reh-ah BROO-tah)*
Ground bond	**Cable de enlace** *(KAH-bleh deh ehn-LAH-seh)*
Ground connection	**Conexión a tierra** *(koh-nek-SYOHN ah TYEH-rra)*
Ground level	**Planta baja** *(PLAHN-tah BAH-hah)*
Ground wire	**Cable a tierra** *(KAH-bleh ah TYEH-rra)*
Grout	**Lechada de cemento** *(leh-CHAH-dah deh seh-MEHN-toh)*

	Mortero de cemento *(mohr-TEH-roh deh seh-MEHN-toh)*
Guardrail	**Baranda** *(bah-RAHN-dah)*
Guest	**Huésped** *(WEHS-pehd)*
Guest room	**Cuarto de huéspedes** *(KWAR-toh deh WEHS-peh-dehs)*
Guide rail	**Riel de guía** *(RYEHL deh GY-ah)*
Gusset plate	**Placa de unión** *(PLAH-kah deh oon-YOHN)* **Placa de cartela** *(PLAH-kah deh kahr-TEH-lah)*
Gutter	**Gotera** *(goh-TEH-rah)* **Canal** *(kah-NAHL)* **Canaleta** *(kah-nah-LEH-tah)*
Gypsum	**Yeso** *(YEH-soh)*
Gypsum board	**Panel de yeso** *(pah-NEHL deh YEH-soh)* **Tablero de yeso** *(tah-BLEH-roh deh YEH-soh)* **Plancha de yeso** *(PLAHN-chah deh YEH-soh)* **Plafón de yeso** *(plah-FOHN deh YEH-soh)*
Gypsum lath	**Listón yesero** *(lees-TOHN yeh-SEH-roh)*
Gypsum plaster	**Revoque de yeso** *(reh-VOH-keh deh YEH-soh)*
Gypsum wallboard	**Panel de yeso** *(pah-NEHL deh YEH-soh)* **Plancha de yeso** *(PLAHN-cha deh YEH-soh)*

Hallway
Pasillo
(pah-SEE-joh)

Handicapped
Discapacitado
(dees-kah-pah-see-TAH-doh)

Minusválido
(mee-noos-VAH-lee-doh)

Handle
Manipular
(mah-nee-poo-LAHR)

Manija
(mah-NEE-hah)

Mango
(MAHN-goh)

Brazo
(BRAH-soh)

Agarradera
(ah-gah-rrah-DEH-rahs)

Handling
Manipulación
(mah-nee-poo-lah-SYOHN)

Handrail
Pasamanos
(pah-sah-MAH-nohs)

Hangers
Ganchos
(GAHN-chohs)

Colgaderos
(kohl-gah-DEH-rohs)

Hanger's saddle
Montura de la percha
(mon-TOO-rah deh lah PEHR-chah)

Hangings
Cortinajes
(kor-tee-NAH-hehs)

Colgaderos
(kohl-gah-DFH-rohs)

Hardboard
Tablero duro
(tah-BLEH-roh DOO-roh)

Hatch
Compuerta
(kohm-PWEHR-tah)

Haunches	**Cartelas** *(kahr-TEH-lahs)*
Hazard	**Peligro** *(peh-LEE-groh)*
Hazardous	**Peligroso** *(peh-lee-GROH-soh)*
	Nocivo *(noh-SEE-voh)*
	Dañino *(dah-NYEE-noh)*
Head (door frame)	**Dintel (de la puerta)** *[deen-TEHL (deh lah PWEHR-tah)]*
Head joint	**Junta vertical** *(HOON-tah vehr-tee-KAHL)*
Header	**Cabezal** *(kah-beh-SAL)*
	Dintel *(deen-TEHL)*
Heater	**Calefactor** *(kahl-eh-fak-TOHR)*
	Estufa *(ehs-TOO-fah)*
	Calentador *(kahl-ehn-tah-DOHR)*
Heating	**Calefacción** *(kahl-eh-fak-SYOHN)*
High rise building	**Edificio de gran altura** *(eh-dee-FEE-syoh deh grahn ahl-TOO-rah)*
Highly toxic material	**Material altamente tóxico** *(mah-teh-RYAHL AHL-tah-mehn-teh TOHC-see-koh)*
High-piled storage	**Almacenamiento en pilas altas** *(ahl-mah-seh-nah-MYEHN-toh ehn PEE-lahs AHL-tahs)*
Hinge	**Bisagra** *(bee-SAH-grah)*
Hip	**Lima** *(LEE-mah)*
	Lima hoya *(LEE-mah OH-yah)*

	Lima tesa *(LEE-mah TEH-sah)*
Hip roof	**Techo a cuatro aguas** *(TEH-cho ah KWAH-tro AH-gwas)*
Hip tile	**Teja para limas** *(TEH-hah PAH-rah LEE-mahs)*
Hod	**Cuezo** *(KWEH-soh)*
Hold-down anchor	**Ancla de retención** *(AHN-klah deh reh-tehn-SYOHN)*
Hole	**Hoyo** *(OH-yoh)*
	Agujero *(ah-goo-HEH-roh)*
	Boquete *(boh-KEH-teh)*
Hood (chimney/kitchen)	**Campana (chimenea/cocina)** *[kahm-PAH-nah (chee-meh-NEH-ah/koh-SEE-nah)]*
Hoop	**Lazo** *(LAH-soh)*
Horizontal bracing system	**Sistema de arriostramiento horizontal** *(sees-TEH-mah deh ah-rryoh-strah-MYEHN-toh oh-ree-sohn-TAHL)*
Horizontal pipe	**Tubo horizontal** *(TOO-boh oh-ree-sohn-TAHL)*
Hose bibb valves	**Válvulas para grifos de mangueras** *(VAHL-voo-lahs PAH-rah GREE-fohs deh mahn-GEH-rahs)*
Hose	**Manguera** *(mahn-GEH-rah)*
Hose threads	**Rosca de manguera** *(ROHS-kah deh mahn-GEH-rah)*
Hot-dipped	**Inmerso en caliente** *(een-MEHR-soh ehn kah-LYEHN-teh)*

Hot bus bar	**Bandeja de carga** *(bahn-DEH-hah deh KAHR-gah)* **Barra ómnibus de carga** *(BAH-rra OHM-nee-boos deh KAHR-gah)*
Hot water	**Agua caliente** *(AH-gwah kah-LYEHN-teh)*
House trap	**Trampa doméstica** *(TRAHM-pah doh-MEHS-tee-kah)*
HVAC	**Calefacción, ventilacion y aire acondicionado** *(kah-leh-fahk-SYOHN, vehn-tee-lah-SYOHN ee AH-ee-reh ah-kohn-dee-syoh-NAH-doh)*
Hydrated lime	**Cal hidratada** *(KAHL ee-drah-TAH-dah)* **Calhidra** *(kah-LEE-dra)*
Hydraulic cement	**Cemento hidráulico** *(seh-MEHN-toh ee-DRAH-oo-lee-koh)*

I

I-Joist	**Vigueta de madera** *(vee-GEH-tah deh ma-DEH-rah)* **Vigueta en doble T** *(vee-GEH-tah ehn DOH-bleh TEH)* **Vigueta-I** *(vee-GEH-tah EE)*
Incline	**Declive** *(deh-KLEE-veh)* **Inclinación** *(een-klee-nah-SYOHN)* **Ladera** *(lah-DEH-rah)* **Pendiente** *(pehn-DYEHN-teh)* **Inclinar** *(een-klee-NAHR)* **Ladear** *(lah-deh-AHR)*
Inspector	**Inspector** *(eens-pek-TOHR)* **Supervisor** *(soo-pehr-vee-SOHR)*
Insulating	**Aislante** *(ah-ees-LAHN-teh)*
Insulation	**Aislamiento** *(ah-ees-lah-MYEHN-toh)* **Aislante** *(ah-ees-LAHN-teh)*
Interior room	**Cuarto interior** *(KWAR-toh een-teh-RYOR)*
Interlay	**Contrachapar** *(kohn-trah-chah-PAHR)*
Interlayment	**Capa intermedia** *(KAH-pah een-tehr-MEH-dee-ah)*

Interlocking

Enclavamiento
(ehn-klah-vah-MYEHN-toh)

**Interlocking
roofing tiles**

Tejas entrelazadas para techo
*(TEH-hahs ehn-treh-lah-SAH-dahs PAH-rah
TEH-choh)*

**International Building
Code**

**Código Internacional de la
Edificación**
*(KOH-dee-goh een-tehr-nah-syoh-NAL deh
lah eh-dee-fee-kah-SYOHN)*

International Fire Code

**Código Internacional de
Protección contra Incendios**
*(KOH-dee-goh een-tehr-nah-syoh-NAL deh
proh-tehk-SYOHN KOHN-trah een-SEHN-
dee-ohs)*

**International Mechanical
Code**

**Código Internacional de
Instalaciones Mecánicas**
*(KOH-dee-goh een-tehr-nah-syoh-NAL deh
een-stah-lah-SYOH-nehs meh-KAH-nee-
kahs)*

**Internacional
Plumbing Code**

**Código Intenacional para
Instalaciones Hidráulicas
y Sanitarias**
*(KOH-dee-goh een-tehr-nah-syoh-NAL
PAH-rah een-stah-lah-SYOH-nehs ee-
DRAH-oo-lee-kahs ee sah-nee-TAH-ryahs)*

Intertied

Entrelazados
(ehn-treh-lah-SAH-dohs)

Intervening rooms

Cuartos intermedios
(KWAR-tohs een-tehr-MEH-dee-ohs)

Isolation joint

Junta de aislamiento
(HOON-tah deh ah-ees-lah-MYEHN-toh)

Jacking force Fuerza
de estiramiento
(FWEHR-sah deh ehs-tee-rah-MYEHN-toh)

Jamb (door frame) Jamba
(HAHM-bah)

Quicial
(kee-SYAHL)

Jobsite Lugar de la obra/en la obra
(loo-GAR deh lah OH-brah/ehn lah OH-brah)

Sitio de construcción
(SEE-tee-oh deh kohn-strook-SYOHN)

Joint Unión
(oon-YOHN)

Junta
(HOON-tah)

Joint compound Pasta de muro
(PAHS tah deh MOO-roh)

Joist Vigueta
(vee-GEH-tah)

Viga
(VEE-gah)

Joist, end Vigueta esquinera
(vee-GEH-tah es-kee-NEH-ra)

Joist, floor Vigueta del piso
(vee-GEH-tah del PEE-soh)

Joist hanger Estribo para vigueta
(ehs-TREE-boh PAH-rah vee-GEH-ta)

Junction Empalme
(ehm-PAHL-meh)

Unión
(oon-YOHN)

40

Kettle	**Caldera** *(kahl-DEH-rah)*
Key	**Llave** *(JAH-veh)*
Keystone	**Clave** *(KLAH-veh)*
King post	**Poste principal** *(POHS-teh preen-see-PAL)* **Columna** *(koh-LOOM-nah)*
Kiosk	**Quiosco** *(KYOHS-koh)*
Kitchen oven	**Horno de cocina** *(OHR-noh deh koh-SEE-nah)*
Kitchen stove	**Estufa de cocina** *(ehs-TOO-fah deh koh-SEE-nah)*
Kitchen sink	**Fregadero de cocina** *(freh-gah-DEH-roh deh koh-SEE-nah)*
Knockout	**Agujero ciego** *(ah-goo-HEH-roh SYEH-goh)*
Kraft paper	**Papel Kraft** *(pah-PEHL krahft)*

Ladder	**Escalera** *(ehs-kah-LEH-rah)*
Landing (stair)	**Descanso de escaleras** *(dehs-KAHN-soh deh ehs-kah-LEH-rahs)*
Lap siding	**Revestimiento de tablas con traslape/solape** *(reh-vehs-tee-MYEHN-toh deh TAH-blahs kohn trahs-LAH-peh/soh-LAH-peh)*
Lap splice	**Traslape** *(trahs-LAH-peh)*
Lapping	**Traslapo** *(trahs-LAH-poh)*
Latch	**Cerrojo** *(seh-RROH-hoh)*
Latching device	**Dispositivo de traba** *(dees-poh-see-TEE-voh deh TRAH-bah)*
Lateral (pipe)	**Ramal lateral** *(rahm-AHL lah-teh-RAHL)*
Lath	**Listón** *(lees-TOHN)* **Malla de enlucir** *(MAH-jah deh ehn-loo-SEER)* **Tiras de yeso** *(TEE-rahs deh YEH-soh)* **Tira** *(TEE-rah)*
Lawn	**Césped** *(SEHS-pehd)* **Pasto** *(PAHS-toh)*
Lay out	**Croquis** *(KROH-kees)*

	Diseño *(dee-SEH-nyo)*
Leader (pipe)	**Tubo de bajada** *(TOO-boh deh bah-HAH-dah)*
Lead-free	**Sin plomo** *(seen PLOH-moh)*
Lead-free **solder and flux**	**Soldadura y fundente** **sin plomo** *(sohl-dah-DOO-rah ee foon-DEN-teh seen* *PLOH-moh)*
Ledger	**Travesaño** *(tra-veh-SAH-nyo)*
	Solera *(soh-LEH-rah)*
Lever	**Palanca** *(pah-LAHN-kah)*
Lift	**Levantamiento** *(leh-vahn-tah-MYEHN-toh)*
	Alzada *(ahl-SAH-dah)*
	Lote *(LOH-teh)*
	Hormigonada *(ohr-mee-goh-NAH-dah)*
	Colada *(koh-LAH-dah)*
Lightbulbs	**Focos** *(FOH-kohs)*
	Bombillas *(bohm-BEE-jahs)*
Light fixture	**Artefacto de iluminación** *(ahr-teh-FAHK-toh deh ee-loo-mee-nah-* *SYOHN)*
Limbs	**Miembros** *(MYEHM-brohs)*
	Extremidades *(ehk-streh-mee-DAH-dehs)*
Lime putty	**Mastique de cal** *(mah-STEE-keh deh kahl)*
Limestone	**Caliza** *(kah-LEE-sah)*

Line and grade	**Trazar y nivelar** *(trah-SAR ee nee-veh-LAR)*
Lining	**Recubrimiento** *(reh-koo-bree-MYEHN-toh)* **Revestimiento** *(reh-vehs-tee-MYEHN-toh)*
Link, Linkage	**Enlace** *(ehn-LAH-seh)* **Tirante** *(tee-RAHN-teh)* **Conexión** *(koh-nek-SYOHN)*
Lintel	**Dintel** *(deen-TEHL)*
Live load	**Cargas vivas** *(KAHR-gahs VEE-vahs)* **Carga variable** *(KAHR-gah vah-ree-AH-bleh)*
Load combination	**Combinación de cargas** *(kohm-bee-nah-SYOHN deh KAHR-gahs)*
Load-bearing joist	**Viga de carga** *(VEE-gah deh KAHR-gah)*
Loaded area	**Área cargada** *(AH-reh-ah kahr-GAH-dah)* **Área sometida a carga** *(AH-reh-ah soh-meh-TEE-dah ah KAHR-gah)*
Lobby	**Lobby** *(LOH-bee)* **Vestíbulo** *(vehs-TEE-boo-loh)*
Local vent stack	**Tubo vertical de ventilación** *(TOO-boh vehr-tee-KAHL deh vehn-tee-lah-SYOHN)*
Lock	**Candado** *(kan-DAH-doh)* **Cerradura** *(seh-rrah-DOO-rah)* **Cerrojo** *(seh-RROH-hoh)*
Lock bolts	**Pernos de seguridad** *(PEHR-nohs deh seh-goo-ree-DAHD)*

Locking receptacle	**Tomas con traba** *(TOH-mahs kohn TRAH-bah)*
Lodging house	**Hospedaje** *(ohs-peh-DAH-heh)*
Loop	**Lazadas** *(lah-SAH-dahs)*
Lot	**Terreno** *(teh-<u>RR</u>EH-noh)*
	Lote *(LOH-teh)*
Louver	**Celosía** *(seh-loh-SEE-ah)*
Lumber	**Madera de construcción** *(mah-DEH-rah deh kohns-trook-SYOHN)*
	Madera elaborada *(mah-DEH-rah eh-lah-boh-RAH-dah)*
LVL beams	**Vigas de madera microlaminada** *(VEE-gahs deh mah-DEH-rah mee-croh-lah-mee-NAH-dah)*

Mail box	**Buzón** *(boo-SOHN)*
Main	**Principal** *(preen-see-PAHL)*
	Matriz *(mah-TREES)*
Main breaker	**Interruptor automático principal** *(een-teh-rroop-TOHR ah-oo-toh-MAH-tee-koh preen-see-PAHL)*
Main power cable	**Cable principal** *(KAH-bleh preen-see-PAHL)*
Main vent	**Respiradero matriz** *(rehs-pee-rah-DEH-roh mah-TREES)*
Mall	**Centro comercial** *(SEHN-troh koh-mehr-SYAHL)*
Manager	**Gerente** *(heh-REHN-teh)*
Manhole	**Pozo de confluencia** *(POH-soh deh kohn-FLWEHN-syah)*
	Boca de inspección *(BOH-kah deh eens-pehk-SYOHN)*
	Boca de acceso *(BOH-kah deh ahk-SEH-soh)*
	Pozo de entrada *(POH-soh deh ehn-TRAH-dah)*
Mansard roof	**Mansarda** *(mahn-SAHR-dah)*
Mansion	**Mansión** *(mahn-SYOHN)*
	Residencia *(reh-see-DEN-syah)*

Manual pull station   **Alarma de incendio manual**
*(ah-LAHR-mah deh een-SEHN-dee-oh
mah-NWAHL)*

Marquee   **Marquesina**
(mar-keh-SEE-nah)

Mason   **Albañil**
(ahl-bah-NYEEL)

Masonry   **Mampostería**
(mahm-pohs-teh-REE-ah)

                             Albañilería
(ahl-bah-nyee-leh-REE-ah)

Mastic   **Mastique**
(mahs-TEE-keh)

Means of egress   **Medios de salida**
(MEH-dee-ohs deh sah-LEE-dah)

Measuring tape   **Cinta de medir**
(SEEN-tah deh meh-DEER)

Mechanical anchorage   **Anclaje mecánico**
(ahn-KLAH-heh meh-KAH-nee-koh)

Metal deck   **Plataforma metálica**
(plah-tah-FOR-mah meh-TAH-lee-kah)

Metal flagpole   **Mástil metálico**
(MAHS-teel meh-TAH-lee-koh)

Metal roof covering   **Cubierta metálica para techos**
*(koo-BYEHR-tah meh-TAH-lee-kah PAH-rah
TEH-chohs)*

Metal scribe   **Trazador de metal**
(trah-sah-DOHR deh meh-TAHL)

Meter   **Medidor**
(meh-dee-DOHR)

Mixing valve   **Llave de mezcla**
(JAH-veh deh MEHS-clah)

Moist curing   **Curado con humedad**
(koo-RAH-doh kohn oo-meh-DAHD)

Molding   **Moldura**
(mohl-DOO-rah)

Mortar   **Mortero**
(mohr-TEH-roh)

Argamasa
(ar-gah-MAH-sah)

Mezcla
(MEHS-klah)

Mortise   **Ranura**
                                (rah-NOO-rah)

Mudsill   **Solera de base**
                                (soh-LEH-rah deh BAH-seh)

Mullion (door)   **Larguero central**
                                (lar-GEH-roh sehn-TRAHL)

Multiple gabled roofs   **Techos a aguas múltiples**
                                (TEH-chohs ah AH-gwahs MOOL-tee-plehs)

Nailing, face	**Con clavos sumidos** *(kohn KLAH-vohs soo-MEE-dohs)*
Nailing strip	**Listón para clavar** *(lee-STOHN PAH-rah klah-VAHR)*
Nails	**Clavos** *(KLAH-vohs)*
Nameplate	**Placa de características** *(PLAH-kah deh kahr-ahk-teh-REES-tee-kahs)* **Placa de datos** *(PLAH-kah deh DAH-tohs)* **Placa de fábrica** *(PLAH-kah deh FAH-bree-kah)*
Natural ventilation	**Ventilación natural** *(vehn-tee-lah-SYOHN nah-too-RAHL)*
Neutral service wire	**Cable principal neutro** *(KAH-bleh preen-see-PAHL NEHOO-troh)*
Neutral wire	**Cable neutro** *(KAH-bleh NEHOO-troh)*
Nonabrasive/abrasive materials	**Materiales no abrasivos /abrasivos** *(mah-teh-RYAH-lehs noh ah-brah-SEE-vohs/ah-brah-SEE-vohs)*
Noncombustible materials	**Materiales no combustibles** *(mah-teh-RYAH-lehs noh kohm-boos-TEE-blehs)*
Nonwoven	**No tejido** *(noh teh-HEE-doh)*
Nosings	**Vuelos** *(VWEH-lohs)*

Notch

Muesca
(MWEHS-kah)

Corte
(KOHR-teh)

Incisión
(een-see-SYOHN)

Nuisance

Perjuicio
(pehr-hoo-EE-syoh)

Nursing homes

Asilos de ancianos
(ah-SEE-lohs deh ahn-SYAH-nohs)

Ancianatos
(ahn-syah-NAH-tohs)

Casas de convalecencia
(KAH-sahs deh kohn-vah-leh-SEHN-syah)

Nut

Tuerca
(TWEHR-kah)

O

Occupancy	**Destino** *(deh-STEE-noh)*
	Tenencia *(teh-NEHN-syah)*
	Actividades *(ahk-tee-vee-DAH-dehs)*
	Clasificación *(klah-see-fee-kah-SYOHN)*
	Función *(foon-SYOHN)*
	Ocupación *(oh-koo-pah-SYOHN)*
	Zona *(soh-NAH)*
	Capacidad *(kah-pah-see-DAHD)*
Occupant load	**Número de ocupantes** *(NOO-meh-roh deh oh-koo-PAHN-tehs)*
Offset	**Desplazamiento** *(dehs-plah-sah-MYEHN-toh)*
	Pieza en "S" *(PYEH-sah ehn "EH-seh")*
	Pieza de inflexión *(PYEH-sah deh een-flek-SYOHN)*
	Desvío *(dehs-VEE-oh)*
	Compensar *(kohm-pehn-SAR)*
Offset bars	**Barras desviadas** *(BAH-rrahs dehs-VYAH-dahs)*
On center	**De centro a centro** *(deh SEHN-troh ah SEHN-troh)*
Open air	**Aire libre** *(AH-ee-reh LEE-breh)*

Opening	**Abertura** *(ah-behr-TOO-rah)*
OSB (oriented strand board)	**Tableros de viruta orientada** *(tah-BLEH-rohs deh vee-ROO-tah oh-ryehn-TAH-dah)*
Outlet box	**Caja de enchufe** *(KAH-hah deh ehn-CHOO-feh)*
	Caja de tomacorriente *(KAH-hah deh toh-mah-koh-<u>RR</u>YEHN-teh)*
Overhang	**Voladizo** *(voh-lah-DEE-so)*
	Vuelo *(VWEH-loh)*
	Alero *(ah-LEH-roh)*
Overhaul	**Reparación** *(reh-par-ah-SYOHN)*
	Reparo *(reh-PAH-roh)*
Overlap	**Traslape** *(trahs-LAH-peh)*
	Sobresolape *(soh-breh-soh-LAH-peh)*
	Superposición *(soo-pehr-poh-see-SYOHN)*
Override	**Cancelar** *(kahn-seh-LAR)*
	Anular *(ah-noo-LAR)*
Overstrength	**Sobreresistencia** *(soh-breh-reh-sees-TEHN-syah)*
Overturning	**Volcamiento** *(vohl-kah-MYEHN-toh)*
	Vuelco *(VWEHL-koh)*
	Volteo *(vohl-TEH-oh)*
Oxidizers	**Oxidantes** *(ohk-see-DAHN-tehs)*

Pallet	**Estante** *(ehs-TAHN-teh)* **Tarima** *(tah-REE-mah)* **Plataforma de carga** *(plah-tah-FOHR-mah deh CAHR-gah)*
Paint	**Pintura** *(peen-TOO-rah)*
Painter	**Pintor** *(peen-TOHR)*
Panel edge clips	**Pinzas de canto de panel** *(peen-sahs deh KAHN-toh deh pah-NEHL)*
Panel sheathing	**Revestimiento de tableros** *(reh-veh-stee-MYEHN-toh deh tah-BLEH-rohs)*
Panel zone	**Franja de tableros** *(FRAHN-hah deh tah-BLEH-rohs)*
Paneling	**Empanelado** *(ehm-pah-neh-LAH-doh)*
Panic bar	**Barra de emergencia** *(BAH-rrah deh eh-mehr-HEN-syah)*
Panic hardware	**Herrajes antipánico** *(eh-RRAH-hehs ahn-tee-PAH-nee-koh)* **Cerrajería o herrajes de emergencia** *(seh-rrah-heh-REE-ah oh eh-RRAH-hehs deh) (eh-mehr-HEN-syah)*
Paper dispensers	**Dispensadores de papel** *(dees-pehn-sah-DOH-rehs deh pah-PEHL)*
Parapet wall	**Muro de parapeto** *(MOO-roh deh pah-rah-PEH-toh)*
Particleboard	**Madera aglomerada** *(mah-DEH-rah ah-gloh-meh-RAH-dah)*

53

Partition

Tabique
(tah-BEE-keh)

Separación
(seh-pah-rah-SYOHN)

División
(dee-vee-SYOHN)

Partition, folding

Tabique plegable
(tah-BEE-keh pleh-GAH-bleh)

Partition, movable

Tabique movible
(tah-BEE-keh moh-VEE-bleh)

Partition, portable

Tabique portátil
(tah-BEE-keh pohr-TAH-teel)

Passageway

Pasillo
(pah-SEE-joh)

**Passageway
(chimney)**

Conducto de humo
(kohn-DOOK-toh deh OO-moh)

Pavement

Pavimento
(pah-vee-MEHN-toh)

Pedestrian walkway

Camino peatonal
(kah-MEE-noh peh-ah-toh-NAHL)

Penthouse

Cuarto de azotea
(KWAR-toh deh ah-soh-TEH-ah)

Perimeter foundation

Fundación perimetral
(foon-dah-SYOHN peh-ree-meh-TRAHL)

Performance

Desempeño
(deh-sehm-PEH-nyoh)

Rendimiento
(rehn-dee-MYEHN-toh)

Comportamiento
(kohm-pohr-tah-MYEHN-toh)

Perlite

Perlita
(pehr-LEE-tah)

Permit

Permiso (de construcción)
[pehr-MEE-soh (deh kohns-trook-SYOHN)]

Pier

Pilar
(pee-LAHR)

Pilón
(pee-LOHN)

Machón
(mah-CHOHN)

	Dado *(DAH-doh)*
Piles	**Pilotes** *(pee-LOH-tehs)*
Pipe, Piping	**Cañería** *(kah-nyeh-REE-ah)*
	Caño *(KAH-nyoh)*
	Tubería *(too-beh-REE-ah)*
	Tubo *(TOO-boh)*
Plan review/reviewer	**Revisión/Revisor de planos** *(reh-vee-SYOHN/reh-vee-SOHR deh PLAH-nohs)*
Plank	**Tablón** *(tah-BLON)*
Planking	**Entablonado** *(ehn-tah-bloh-NAH-doh)*
	Tablones *(tah-BLOH-nehs)*
Plaster	**Azotado** *(ah-soh-TAH-doh)*
	Jaharro *(hah-AH-rroh)*
	Enjarre *(ehn-HAH-rreh)*
	Enlucido *(ehn-loo-SEE-doh)*
	Yeso *(YEH-soh)*
Plaster backing	**Soporte para forjados** *(soh-POHR-teh PAH-rah fohr-HAH-dohs)*
Plastering	**Revoque** *(reh-VOH-keh)*
	Enlucido *(ehn-loo-SEE-doh)*
	Repello *(reh-PEH-joh)*
	Forjados *(fohr-HAH-dohs)*
Plastic foam	**Espuma de plástico** *(ehs-POO-mah deh PLAH-stee-koh)*

Plastic insulator	**Aislante plástico** *(ah-ee-SLAHN-teh PLAH-stee-koh)*
Plate girder	**Viga de alma llena** *(VEE-gah deh AHL-mah JEH-nah)*
Plenum	**Pleno** *(PLEH-noh)* **Cámara de distribución de aire** *(KAH-mah-rah deh dees-tree-boo-SYOHN deh AH-ee-reh)*
Plug	**Clavija** *(klah-VEE-hah)* **Enchufe** *(ehn-CHOO-feh)*
Plug fuse	**Fusible de rosca** *(foo-SEE-bleh deh ROHS-kah)*
Plumber	**Plomero** *(ploh-MEH-roh)*
Plumbing	**Instalaciones hidráulicas y sanitarias** *(een-stahl-ah-SYOHN-ehs ee-DRAH-oo-lee-kahs ee sah-nee-TAH-ryahs)* **Plomería** *(ploh-meh-REE-ah)*
Plumbing appliance	**Mueble sanitario** *(MWEH-bleh sah-nee-TAH-ryoh)*
Plumbing fixture	**Artefacto sanitario** *(ahr-teh-FAHK-toh sah-nee-TAH-ryoh)*
Plunger	**Destapacaños** *(dehs-tah-pah-CAH-nyos)* **Sopapa** *(soh-PAH-pah)*
Ply	**Capa** *(KAH-pah)*
Plywood	**Madera prensada** *(mah-DEH-rah prehn-SAH-dah)* **Tableros de madera prensada** *(tah-BLEH-rohs deh mah-DEH-rah prehn-SAH-dah)*
Point load	**Carga puntal** *(KAHR-gah poon-TAHL)*

Poles, Posts

Postes
(POHS-tehs)

Pollution

Polución
(poh-loo-SYOHN)

Ponding

Estancamiento de agua
(ehs-tahn-kah-MYEHN-toh deh AH-gwah)

Portable

Portátil
(pohr-TAH-teel)

Pour coat

Capa de colada
(KAH-pah deh koh-LAH-dah)

Capa de vaciado
(KAH-pah deh vah-SYAH-doh)

Power doors

Puertas mecánicas
(PWEHR-tahs meh-KAH-nee-kahs)

Power strip

Zapatilla eléctrica
(sah-pah-TEE-jah eh-LEHK-tree-kah)

Power supply

Fuente de alimentación
(FWEHN-teh deh ah-lee-mehn-tah-SYOHN)

Premises

Local
(loh-KAHL)

Sitio
(SEE-tee-oh)

Press box

Palco de prensa
(PAHL-koh deh PREHN-sah)

Pressure

Presión
(preh-SYOHN)

Prestressed concrete

Hormigón preesforzado
(ohr-mee-GOHN preh-ehs-fohr-SAH-doh)

Hormigón precargado
(ohr-mee-GOHN preh-kahr-GAH-doh)

Hormigón precomprimido
(ohr-mee-GOHN preh-kohm-pree-MEE-doh)

Hormigón prefatigado
(ohr-mee-GOHN preh-fah-tee-GAH-doh)

Primed

Imprimado
(eem-pree-MAH-doh)

Primer

Imprimador
(eem-pree-mah-DOHR)

Private

Privado
(pree-VAH-doh)

Property	**Propiedad** *(proh-pyeh-DAHD)* **Parcela** *(par-SEH-lah)*
Property line	**Línea de propiedad** *(LEE-neh-ah deh proh-pyeh-DAHD)* **Lindero** *(leen-DEH-roh)*
Proportion	**Proporción** *(proh-pohr-SYOHN)* **Proporcionalidad** *(proh-pohr-syoh-nah-lee-DAHD)* **Dimensionar** *(dee-mehn-syoh-NAHR)* **Determinar las dimensiones** *(deh-tehr-mee-NAHR lahs dee-meh-SYOH-nehs)*
Proportioned	**Dimensionado** *(dee-mehn-syoh-NAH-doh)*
Provision/Proviso	**Disposición** *(dee-spoh-see-SYOHN)* **Estipulación** *(ehs-tee-poo-lah-SYOHN)*
Public safety	**Seguridad pública** *(seh-goo-ree-DAHD POO-blee-kah)* **Protección al público** *(proh-tehk-SYOHN ahl POO-blee-koh)*
Public way	**Vía pública** *(VEE-ah POO-blee-kah)*
Public welfare	**Bienestar público** *(byehn-ehs-TAHR POO-blee-koh)*
Pump	**Bomba** *(BOHM-bah)*
Putty coat	**Enlucido** *(ehn-loo-SEE-doh)*

Quarry	**Cantera** *(kahn-TEH-ra)*
Queen post	**Columna** *(koh-LOOM-nah)*
Quick-acting bolt	**Perno de acción rápida** *(PEHR-noh deh ahk-SYOHN RRAH-pee-dah)*
Quick-acting coupler	**Acoplamiento de acción rápida** *(ah-koh-plah-MYEHN-toh deh ahk-SYOHN RRAH-pee-dah)*
Quick-acting nut	**Tuerca de acción rápida** *(TWEHR-kah deh ahk-SYOHN RRAH-pee-dah)*
Quicklime	**Cal viva** *(KAHL vee-vah)*
Quick setting	**De fraguado/ endurecimiento rápido** *(deh frah-GWAh-doh/ehn-doo-reh-see-MYEHN-toh RRAH-pee-doh*
Quoin	**Piedra angular** *(PYEH-drah ahn-goo-LAHR)* **Cuña** *(KOO-nyah)*
Quota	**Cuota** *(kwoh-tah)*

59

Rabbet	**Muesca** *(MWEHS-kah)* **Ranura** *(rah-NOO-rah)*
Raceways	**Conducto eléctrico** *(kohn-DOOK-toh eh-LEHK-tree-koh)*
Rack	**Cremallera** *(krehm-ah-JEH-rah)* **Tarima** *(tah-REE-mah)*
Rafter	**Cabrio** *(KAH-bree-oh)* **Cabio** *(KAH-bee-oh)*
Rafter ends	**Orillas de cabios** *(oh-REE-yahs deh KAH-bee-ohs)*
Rail	**Cremallera** *(krehm-ah-JEH-rah)* **Baranda** *(bah-RAHN-dah)* **Barandilla** *(bah-rahn-DEE-jah)*
Railing	**Baranda** *(bah-RAHN-dah)* **Barra** *(BAH-_rr_ah)* **Carril** *(kah-_RREEL_)*
Range poweroutlet	**Tomacorriente/Enchufe para estufa** *(toh-mah-koh-_RR_YEHN-teh/ehn-CHOO-feh PAH-rah ehs-TOO-fah)*
Rate	**Relación** *(reh-lah-SYOHN)*

	Proporción *(proh-pohr-SYOHN)*
	Razón *(rah-SOHN)*
Rating	**Clasificación** *(klah-see-fee-kah-SYOHN)*
Ratio	**Relación** *(reh-lah-SYOHN)*
	Cociente *(koh-SYEHN-teh)*
	Razón *(rah-SON)*
	Coeficiente *(koh-eh-fee-SYEHN-teh)*
Ready access	**Acceso libre** *(ahk-SEH-soh LEE-breh)*
Rebar	**Barra de refuerzo** *(BAH-rrah de reh-FWEHR-soh)*
	Varilla *(vah-REE-jah)*
Redwood	**Madera de secoya** *(mah-DEH-rah deh seh-KOH-yah)*
Reformatory	**Reformatorio** *(reh-fohr-mah-TOH-ree-oh)*
Refuge area	**Área de refugio** *(AH-reh-ah deh reh-FOO-hee-oh)*
Region	**Región** *(reh-hee-ONH)*
	Tramo *(TRAH-moh)*
Register (baseboard/ ceiling/wall)	**Rejilla (de piso/techo/pared)** *[reh-HEE-jah (deh pee-soh/TEH-cho/pa-REHD)]*
Reglet	**Regleta** *(Reh-GLEH-tah)*
Regulator	**Regulador** *(reh-goo-lah-DOHR)*
Reinforced masonry	**Mampostería reforzada** *(mahm-pohs-teh-REE-ah reh-fohr-SAH-dah)*

61

Reinforcement	**Refuerzo** *(reh-FWEHR-soh)*
	Armadura *(ar-mah-DOO-rah)*
Release	**Descarga** *(dehs-KAHR-gah)*
	Liberación *(lee-beh-rah-SYOHN)*
	Desenganchador *(deh-sehn-gahn-chah-DOHR)*
	Desenganchar *(deh-sehn-gahn-CHAR)*
Relief valve	**Válvula de alivio** *(VAHL-voo-lah deh ah-LEE-vee-oh)*
	Llave de alivio *(JAH-veh deh ah-LEE-vee-oh)*
Removal	**Eliminación** *(eh-lee-mee-nah-SYOHN)*
	Remoción *(reh-moh-SYOHN)*
Repair	**Reparación** *(reh-par-ah-SYOHN)*
Reports	**Informes** *(een-FOR-mehs)*
	Reportes *(reh-POHR-tehs)*
Reshores	**Puntales de refuerzo** *(poon-TAH-lehs deh reh-FWEHR-zoh)*
Residence	**Residencia** *(reh-see-DEN-syah)*
Restraints	**Sujetadores** *(soo-heh-tah-DOH-rehs)*
	Fijadores *(fee-hah-DOH-rehs)*
Restroom	**Baño** *(BAH-nyo)*
	Sanitario *(sah-nee-TAH-ryoh)*
Retrofitting	**Retroajuste** *(reh-troh-ah-HOOS-teh)*

Return bend vent pipe	**Tubo de ventilación con codo doble** *(TOO-boh deh vehn-tee-lah-SYOHN kohn KOH-doh DOH-bleh)*
Return lip	**Remate de borde** *(reh-MAH-teh deh BOR-deh)*
Revolving door	**Puerta giratoria** *(PWEHR-tah hee-rah-TOH-ree-ah)*
Rib	**Costilla** *(koh-STEE-jah)*
Ridge	**Cresta** *(KREH-stah)*
	Cumbrera *(koom-BREH-rah)*
Ridge board	**Tabla de cumbrera** *(TAH-blah deh koom-BREH-rah)*
Ridge tile	**Tejas para cumbrera** *(TEH-hahs PAH-rah koom-BREH-rah)*
Riffled	**Ranurado** *(rah-noo-RAH-doh)*
	Acanalada *(ah-kahn-ah-LAH-dah)*
Rim	**Borde** *(BOR-deh)*
Rim board	**Viga perimetral** *(VEE-gah peh-ree-meh-TRAHL)*
Ring shank nail	**Clavos con fuste corrugado** *(KLAH-vohs kohn FOO-steh koh-rroo-GAH-doh)*
Ringed shanks	**Varillas en aro** *(vah-REE-jahs ehn AR-oh)*
Riser (pipe)	**Tubo vertical** *(TOO-boh vehrt-ee-KAHL)*
Riser (stair)	**Contrahuella** *(kohn-trah-WEH-jah)*
Rivet	**Remache** *(reh-MAH-cheh)*
Rock	**Roca** *(ROH-kah)*

	Piedra *(PYEH-drah)*
Roof	**Techo** *(TEH-choh)*
Roof covering	**Revestimiento de techo** *(reh-veh-stee-MYEHN-toh deh TEH-choh)*
	Cubierta de azotea *(koo-BYEHR-tah deh ah-soh-TEH-ah)*
	Cubierta de techo *(koo-BYEHR-tah deh TEH-choh)*
Roof deck	**Cubierta de techo** *(koo-BYEHR-tah deh TEH-choh)*
Roof drain	**Desagüe de techo** *(deh-SAH-gweh deh TEH-choh)*
Roof, flat	**Techo plano** *(TEH-choh PLAH-noh)*
Roof sheating	**Entarimado de tejado** *(ehn-tah-ree-MAH-doh deh teh-JAH-doh)*
Roof, sloped	**Techo en pendiente** *(TEH-choh ehn pehn-DYEHN-teh)*
Roof tile	**Teja** *(TEH-ha)*
Roofing	**Techado** *(Teh-CHAH-doh)*
Roofing square	**Cuadro de cubierta de techo** *(KWAH-droh deh koo-bee-EHR-tah deh TEH-choh)*
Room	**Cuarto** *(KWAR-toh)*
	Sala *(SAH-lah)*
	Habitación *(ah-bee-tah-SYOHN)*
Room, assembly	**Sala** *(SAH-lah)*
	Salón *(sah-LOHN)*
	Cuarto de asambleas *(KWAR-toh deh ah-sahm-BLEH-ahs)*

Rough-in	**Instalación en obra negra/gruesa** *(een-stah-lah-SYOHN ehn OH-brah NEH-grah/GRWEH-sah)*
Row	**Fila** *(FEE-lah)*
Rubbish	**Basura** *(bah-SOO-rah)*
Rubble	**Escombro** *(ehs-COM-broh)*
Runners	**Largueros** *(lar-GUEH-rohs)*

Safety glazing	**Vidriado de seguridad** *(vee-DRYAH-doh deh seh-goo-ree-DAHD)*
Sand	**Arena** *(ah-REH-nah)*
Sandstone	**Areniscas** *(ah-reh-NEES-kahs)* **Piedra arenisca** *(PYEH-drah ah-reh-NEES-kah)*
Sanitation	**Higiene** *(ee-hee-EH-neh)*
Saw kerf	**Corte de sierra** *(KOHR-teh deh SYEH-<u>rr</u>ah)*
Sawn timber	**Maderos aserrados** *(mah-DEH-rohs ah-seh-<u>RR</u>AH-dohs)*
Sawtooth	**Diente de sierra** *(DYEHN-teh deh SYEH-<u>rr</u>ah)*
Scaffold	**Andamio** *(ahn-DAH-myoh)*
Scaffolding	**Andamiaje** *(ahn-dah-MYAH-heh)*
Schedule	**Horario** *(oh-RAH-ree-oh)* **Planilla** *(plah-NEE-yah)*
Scope	**Alcance** *(ahl-KAHN-seh)*
Scouring	**Rozamiento** *(roh-zah-MYEHN-toh)*
Screw	**Tornillo** *(tohr-NEE-joh)*
Screw connector	**Conector con tornillo** *(koh-nehk-tohr kohn tohr-NEE-joh)*

Sealants	**Selladores** *(seh-jah-DOH-rehs)*
Seasoned wood	**Madera estacionada** *(mah-DEH-rah ehs-tah-syoh-NAH-dah)*
Self-closing	**Autocierre** *(ah-oo-toh-SYEH-rreh)*
Self-closing device	**Dispositivo autocerrante** *(dee-spoh-see-TEE-voh ah-oo-toh-seh-RRAHN-teh)* **Dispositivo de cierre mecanizado** *(dee-spoh-see-TEE-voh deh SYEH-rreh meh-KAH-nee-zah-doh)*
Self-closing faucet	**Grifo de cierre automático** *(GREE-foh deh SYEH-rreh ah-oo-toh-MAH-tee-koh)*
Self-drilling screws	**Tornillos autoperforantes** *(tohr-NEE-johs ah-oo-toh-pehr-foh-RAHN-tehs)*
Self-ignition	**Auto-ignición** *(ah-oo-toh-ee-gnee-SYOHN)*
Self-luminous	**Autoluminoso** *(ah-oo-toh-loo-mee-NOH-soh)*
Self-tapping screws	**Tornillos autorroscantes** *(tohr-NEE-johs ah-oo-toh-rrohs-KAHN-tehs)*
Sensitizer	**Sensiblilzador** *(sehn-see-bee-lee-sah-DOHR)*
Set out	**Resaltar** *(reh-sahl-TAR)*
Setback	**Retiro** *(reh-TEE-roh)*
Sewage	**Aguas negras** *(AH-gwahs NEH-grahs)* **Cloacas** *(KLOAH-kahs)* **Residuos cloacales** *(reh-SEE-dwohs kloah-KAH-lehs)*
Sewer	**Cloaca** *(KLOAH-kah)* **Alcantarilla** *(ahl-cahn-tah-REE-jah)*

Shaft	**Recinto** *(reh-SEEN-toh)*
Shake, wood	**Teja de madera** *(TEH-hah deh mah-DEH-rah)* **Ripia** *(REE-pee-ah)*
Shear wall	**Muro cortante** *(MOO-roh kor-TAHN-teh)* **Muro de corte** *(MOO-roh deh KOR-teh)* **Muro sismorresistente** *(MOO-roh sees-moh-<u>rr</u>eh-sees-TEHN-teh)*
Sheathing	**Entablado** *(ehn-tah-BLAH-doh)*
Sheathing edges	**Bordes del entablado** *(BOHR-dehs dehl ehn-tah-BLAH-doh)*
Sheet	**Pliego** *(PLYEH-goh)* **Chapa** *(CHAH-pah)* **Plancha** *(PLAHN-cha)* **Lámina** *(LAH-mee-nah)* **Tablestacado** *(tah-bleh-stah-KAH-doh)*
Sheet metal	**Lámina/Chapa metálica** *(LAH-mee-nah/CHAH-pah meh-TAH-lee-kah)* **Laminado** *(lah-mee-NAH-doh)*
Sheeting	**Laminado** *(lah-mee-NAH-doh)* **Lámina** *(LAH-mee-nah)*
Shelf	**Repisa** *(reh-PEE-sah)*
Shell	**Cáscara** *(KAHS-kah-rah)* **Cubierta** *(koo-BYEHR-tah)*

Shingle	**Teja** *(TEH-jah)*
	Tejamanil *(teh-hah-mah-NEEL)*
Shingle, asphalt	**Teja de asfalto** *(TEH-ha de as-PHAL-to)*
Shingle, wood	**Ripia** *(REE-pee-ah)*
Shiplap	**Traslape** *(trah-SLAH-peh)*
	Rebajo a media madera *(reh-BAH-hoh ah MEH-dyah mah-DEH-rah)*
Shop	**Taller** *(tah-JEHR)*
Shored/Unshored construction	**Construcción apuntalada/ no apuntalada** *(kohn-strook-SYOHN ah-poon-tah-LAH-dah/noh ah-poon-tah-LAH-dah)*
Short blocking	**Bloques cortos** *(BLOH-kehs KOHR-tohs*
Shotcrete	**Hormigón proyectado** *(ohr-mee-GOHN proh-jehk-TAH-doh)*
	Gunita *(goo-NEE-ta)*
Shotcrete structures	**Estructura de hormigón proyectado** *(ehs-trook-TOO-rah deh ohr-mee-GOHN proh-jehk-TAH-doh)*
	Estructuras de gunita *(ehs-trook-TOO-rahs deh goo-NEE-tah)*
Show window	**Vitrina** *(vee-TREE-nah)*
Showcase	**Armario de exhibición** *(ar-MAH-ree-oh deh ehk-see-bee-SYOHN)*
Shower door	**Mampara de ducha** *(mahm-PAH-rah deh DOO-chah)*
Shower stall	**Ducha** *(DOO-chah)*
	Regadera *(reh-gah-DEH-rah)*

69

	Cuarto de regadera *(KWAR-toh deh reh-gah-DEH-rah)*
Showerhead	**Regadera** *(reh-gah-DEH-rah)*
Shrinkage	**Contracción** *(kohn-trahc-SYOHN)*
	Encogimiento *(ehn-koh-hee-MYEN-toh)*
	Reducción *(reh-dook-SYOHN)*
Shutoff valves	**Válvulas de cierre** *(VAHL-voo-lahs deh SYEH-rreh)*
Side-hinged door	**Puerta con bisagras laterales** *(PWEHR-tah kohn bee-SAH-grahs lah-teh-RAH-lehs)*
Sidewalk	**Acera** *(ah-SEH-rah)*
	Vereda *(veh-REH-dah)*
	Banqueta *(bahn-keh-tah)*
Sill	**Soporte** *(soh-POHR-teh)*
Sill cock	**Grifo de manguera** *(GREE-foh deh mahn-GUEH-rah)*
Sill plate	**Solera inferior** *(soh-LEH-rah een-feh-RYOR)*
Single pole breaker	**Interruptor automático unipolar** *(een-teh-rroop-TOHR ah-oo-toh-MAH-tee-koh oo-nee-poh-LAHR)*
Sink	**Lavabo** *(lah-VAH-boh)*
Sink, kitchen	**Fregadero** *(freh-gah-DEH-roh)*
	Pileta de cocina *(pee-LEH-tah deh koh-SEE-na)*
	Tarja de cocina *(TAR-hah deh koh-SEE-na)*
Site	**Sitio** *(SEE-tee-oh)*

Skylight

Tragaluz
(trah-gah-LOOS)

Claraboya
(klar-ah-BOH-yah)

Skyscraper

Rascacielos
(rahs-kah-SYEH-lohs)

Slab

Losa
(LOH-sah)

Slab foundation

Losa de cimentación
(LOH-sah deh see-mehn-tah-SYOHN)

Slags

Escorias
(ehs-KOH-ree-ahs)

Slate shingle

Teja de pizarra
(TEH-hah deh pee-SAH-<u>rr</u>ah)

Sleeper

Traviesa
(trah-VYEH-sah)

Durmiente
(door-MYEHN-teh)

Sleeve

Camisa
(kah-MEE-sah)

Manga
(MAHN-gah)

Sliding doors/windows

**Puertas/ventanas deslizantes/
corredizas**
*(PWEHR-tahs/vehn-TAH-nahs deh-slee-
SAHN-tehs/koh-rreh-DEE-sahs)*

Slope

Pendiente
(pehn-DYEHN-teh)

Talud
(tah-LOOD)

Declive
(deh-KLEE-veh)

Slump

Asentamiento
(ah-sehn-tah-MYEHN-toh)

Smoke

Humo
(OO-moh)

Smoke barrier

Barrera antihumo
(bah-<u>RR</u>EH-rah ahn-tee-OO-moh)

Smoke curtain

Cortina antihumo
(kor-TEE-nah ahn-tee-OO-moh)

71

Smoke density	**Densidad de humo** *(den-see-DAHD deh OO-moh)*
Smoke detector	**Sensor de humo** *(sehn-SOHR deh OO-moh)*
	Detector de humo *(deh-tehk-TOHR deh OO-moh)*
Smoke exhaust system	**Sistema de extracción de humo** *(sees-TEH-mah deh ehks-trahk-SYOHN deh OO-moh)*
Smoke layer	**Capa de humo** *(kah-pah deh OO-moh)*
Smoke tight	**Impermeables al humo** *(eem-pehr-meh-AH-blehs ahl OO-moh)*
Smoke-control zone	**Zona de control de humo** *(SOH-nah deh kohn-TROL deh OO-moh)*
Smoke-detection system	**Sistema de detección de humo** *(sees-TEH-mah deh deh-tehk-SYOHN deh OO-moh)*
Soffit	**Sofito** *(soh-FEE-toh)*
Soft story	**Piso blando** *(PEE-soh BLAHN-doh)*
	Piso flexible *(PEE-soh flek-SEE-bleh)*
Soil pipe	**Tubo bajante de aguas negras** *(TOO-boh bah-JAHN-teh deh AH-gwahs NEH-grahs)*
Soil stack	**Bajante sanitaria** *(bah-HAHN-teh sah-nee-TAH-ryah)*
Sole plate	**Placa de base** *(PLAH-kah deh BAH-seh)*
Solid-sawn wood	**Madera aserrada de una pieza** *(mah-DEH-rah ah-she-<u>RR</u>AH-dah deh oo-nah PYEH-sah)*
Spalling	**Astilladuras** *(ah-stee-jah-DOO-rahs)*
Span	**Luz** *(loos)*

	Vano
	(VAH-noh)
	Claro
	(KLAH-roh)
Spandrel	**Jácena exterior**
	(HAH-seh-nah ehk-steh-RYOHR)
	Tímpano
	(TEEM-pah-noh)
	Enjuta
	(ehn-HOO-tah)
	Muro de relleno
	(MOO-roh deh reh-JEH-noh)
Spark arrester	**Trampa para chispas**
	(TRAHM-pah PAH-rah CHEES-pahs)
Special hazards	**Peligros especiales**
	(peh-LEE-grohs eh-speh-SYAH-lehs)
Spigot	**Llave**
	(JAH-veh)
	Grifo
	(GREE-foh)
	Canilla
	(kah-NEE-jah)
	Espiga
	(ehs-PEE-gah)
Spike	**Clavo especial para madera**
	(KLAH-voh eh-speh-SYAHL PAH-rah mah-DEH-rah)
Spiked	**Clavado**
	(KLAH-vah-doh)
Spiral stairs	**Escaleras de caracol**
	(ehs-kah-LEH-rahs deh kah-rah-KOHL)
Spirals	**Espirales**
	(eh-spee-RAH-lehs)
Spire	**Aguja**
	(ah-GOO-hah)
Splice	**Empalme**
	(ehm-PAHL-meh)
	Traslape
	(trah-SLAH-peh)
	Junta
	(HOON-tah)

	Unión *(oon-YOHN)*
Splice plates	**Planchas de empalme** *(PLAHN-chas deh ehm-PAHL-meh)*
Splined	**Acanalado** *(ah-kah-nah-LAH-doh)*
Spot mopped	**Adherido en secciones** *(ah-deh-REE-doh ehn sehk-SYOH-nehs)*
Spring	**Resorte** *(reh-SOHR-teh)*
Sprinkler	**Rociador** *(roh-syah-DOHR)*
Sprinkler head	**Boquilla de rociador** *(boh-KEE-jah deh roh-syah-DOHR)*
Sprinkler systems (automatic)	**Sistemas de rociadores (automáticos)** *[sees-TEH-mahs deh roh-SYAH-dohrehs ah-oo-toh-MAH-tee-kohs)]*
Squash blocks	**Bloques de transferencia de carga** *(BLOH-kehs deh trans-feh-REHN-syah deh CAHR-ga)*
Stack	**Tubería vertical/bajante** *(too-beh-REE-ah vehr-tee-KAHL/bah-HAHN-teh)* **Tubo vertical de evacuación** *(TOO-boh vehr-tee-KAHL deh eh-vah-kwah-SYOHN)*
Stack vent	**Respiradero de bajante** *(reh-spee-rah-DEH-roh deh bah-HAHN-teh)* **Tubería bajante de respiradero** *(too-beh-REE-ah bah-HAHN-teh- deh reh-spee-rah-DEH-roh)*
Stainless steel	**Acero inoxidable** *(ah-SEH-roh ee-noc-see-DAH-bleh)*
Stairs	**Escaleras** *(ehs-kah-LEH-rahs)*
Stairway	**Escalera** *(ehs-kah-LEH-rah)*

Stairway, enclosed	**Recinto de escaleras** *(reh-SEEN-toh deh ehs-kah-LEH-rahs)*
Stairwells	**Recinto de escaleras** *reh-SEEN-toh deh ehs-kah-LEH-rahs)*
Standpipe	**Columna hidrante** *(koh-LOOM-nah ee-DRAHN-teh)* **Tubería vertical** *(too-beh-REE-ah vehr-tee-KAHL)*
Standpipe system	**Sistema de columna hidrante** *(sees-TEH-mah deh koh-LOOM-nah hee-DRAHN-teh)*
Stands (reviewing)	**Estrados** *(eh-STRAH-dohs)* **Tribunas** *(tree-BOO-nah)* **Gradas** *(GRAH-dahs)*
Steam	**Vapor** *(vah-POHR)*
Steel	**Acero** *(ah-SEH-roh)*
Steel framing	**Estructura en/de acero** *(eh-strook-TOO-rah ehn / deh ah-SEH-roh)*
Steel studs	**Montantes de acero** *(mohn-TAHN-tehs deh ah-SEH-roh)*
Steeple	**Campanario** *(kahm-pah-NAH-ree-oh)*
Steps	**Escalones** *(ehs-kah-LOH-nehs)* **Peldaños** *(pehl-DAH-nyohs)*
Sterilizer	**Esterilizador** *(eh-steh-ree-lee-sah-DOHR)*
Stiffener	**Pieza de refuerzo** *(PYEH-sah deh reh-foo-EHR-soh)* **Refuerzo** *(reh-foo-EHR-soh)* **Atiesador** *(ah-tee-eh-sah-DOH-rehs)*
Stiffness	**Rigidez** *(ree-he-DEHS)*

Stirrups	**Estribos** *(eh-STREE-bohs)*
Stone	**Piedra** *(PYEH-drah)*
	Roca *(ROH-kah)*
Stops (door frame)	**Topes** *(TOH-pehs)*
Storage room	**Cuarto de almacenamiento** *(KWAR-toh deh ahl-mah-seh-nah-MYEHN-toh)*
	Bodega *(boh-DEH-gah)*
	Depósito *(deh-POH-see-toh)*
Store	**Tienda** *(TYEHN-dah)*
Storm drain	**Alcantarilla** *(ahl-kahn-tar-EE-jah)*
Story	**Piso** *(PEE-soh)*
	Planta *(PLAHN-tah)*
Strap	**Fleje** *(FLEH-heh)*
	Cincho *(SEEN-choh)*
Strapping	**Flejes** *(FLEH-hehs)*
Stress	**Esfuerzo** *(ehs-foo-EHR-soh)*
Strip	**Listón** *(lee-STON)*
Stripping	**Tiras metálicas** *(TEE-rahs meh-TAH-lee-kahs)*
Structure	**Estructura** *(eh-strook-TOO-rah)*
Strut	**Puntal** *(poon-TAHL)*

Stucco	**Revoque** *(reh-VOH-keh)*
	Enlucido *(ehn-loo-SEE-doh)*
	Estuco *(ehs-TOO-koh)*
Stud	**Montante** *(mohn-TAHN-teh)*
	Parante *(par-AHN-teh)*
	Barrote *(ba-RROH-teh)*
Stud anchors	**Barras de anclaje** *(BAH-rrahs deh ahn-KLAH-heh)*
Stud bearing wall	**Muro portante con montante** *(MOO-roh pohr-TAHN-teh kohn mohn-TAHN-teh)*
Stud walls	**Muros con montantes** *(MOO-rohs kohn mohn-TAHN-tehs)*
	Paredes de barrotes *(pah-REH-dehs deh bah-RROH-tehs)*
Stud finder	**Buscador de montantes** *(boos-kah-DOHR deh mohn-TAHN-tehs)*
Subfloor	**Contrapiso** *(kohn-trah-PEE-soh)*
	Bajopiso *(bah-ho-PEE-soh)*
	Subpiso *(soob-PEE-soh)*
Subroof	**Base de techo** *(BAH-seh)*
Substrate	**Substrato** *(soob-STRAH-toh)*
Suite (hotel)	**Suite** *(soo-EE-teh)*
Sump	**Sumidero** *(soo-mee-DEH-roh)*
Sump pump	**Bomba de sumidero** *(BOHM-bah deh soo-mee-DEH-roh)*

Sump vent	**Respiradero de sumidero** *(reh-spee-rah-DEH-roh deh soo-mee-DEH-roh)*
Superintendent	**Superintendente** *(soo-pehr-een-tehn-DEN-teh)*
Supervisor	**Supervisor** *(soo-pehr-vee-SOHR)* **Inspector** *(eens-pek-TOHR)*
Supply	**Suministro** *(soo-mee-NEES-troh)* **Abastecimiento** *(ah-bahs-teh-see-MYEHN-toh)* **Alimentacion** *(ah-lee-mehn-tah-SYOHN)*
Support	**Apoyo** *(ah-POH-joh)* **Soporte** *(soh-POHR-teh)*
Support (v)	**Resistir** *(reh-see-STEER)* **Sostener** *(sohs-teh-NEHR)* **Soportar** *(soh-pohr-TAR)*
Suppressor	**Amortiguador** *(ah-mohr-tee-gwah-DOHR)*
Suspended ceiling	**Falso plafón** *(fahl-soh plah-FOHN)* **Cielorraso suspendido** *(syeh-loh-RRAH-soh soos-pehn-DEE-doh)*
Swelling	**Expansión** *(ehk-spahn-SYOHN)* **Hinchazón** *(heen-chah-SOHN)*
Swimming pool	**Piscina** *(pee-SEE-nah)* **Alberca** *(ahl-BEHR-kah)* **Pileta** *(pee-LEH-tah)*

Swinging door   **Puerta pivotante**
                             (PWEHR-tah pee-voh-TAHN-teh)

Switch   **Interruptor**
                             (een-teh-rroop-TOHR)

                             Apagador
                             (ah-pah-gah-DOHR)

Switch plate   **Placa del interruptor**
                             (PLAH-kah deh een-teh-rroop-TOHR)

Tag
Etiqueta
(eh-tee-KEH-tah)

Taping compound
Pasta de muro
(PAHS tah deh MOO-roh)

Tar
Alquitrán
(ahl-kee-TRAHN)

Brea
(BREH-ah)

Chapopote
(chah-poh-POH-teh)

Tar paper
Papel de brea
(pah-PEHL de BREH-ah)

Tarp
Lona
(LOH-nah)

TEE
T, injerto
(T, een-HEHR-toh)

Technician
Técnico
(TEHK-nee-koh)

Templates
Plantillas
(plahn-TEE-jahs)

Temporary
Provisional
(proh-vee-syoh-NAHL)

Tenant
Inquilino
(een-kee-LEE-noh)

Tendons
Tendones
(tehn-DOH-nehs)

Tensores
(tehn-SOH-rohs)

Tension
Tensión
(tehn-SYOHN)

Terminal
Terminal
(tehr-mee-NAL)

Termite   **Termita**
                       (tehr-MEE-tah)

Test   **Ensayo**
                       (ehn-SAH-joh)

                       Prueba
                       (PRWEH-bah)

                       Someter a ensayo/prueba
                       (soh-meh-TEHR ahehn-SAH-joh/PRWEH-bah)

Test apparatus   **Aparato de ensayo**
                       (ah-par-AH-toh deh ehn-SAH-joh)

                       Aparato de prueba
                       (ah-par-AH-toh deh PRWEH-bah)

Texture   **Textura**
                       (tehks-TOO-rah)

Thawing   **Descongelación**
                       (dehs-kohn-heh-lah-SYOHN)

Thread   **Hilo**
                       (EE-loh)

Threshold   **Umbral**
                       (oom-BRAHL)

Tie   **Amarra**
                       (ah-MAH-rrah)

                       Ligadura
                       (lee-gah-DOO-rah)

                       Tirante
                       (tee-RAHN-teh)

Tier   **Hilera**
                       (ee-LEH-rah)

Tile   **Teja**
                       (TEH-hah)

Tile, floor   **Baldosas**
                       (bahl-DOH-sahs)

Tile, masonry   **Ladrillo cerámico**
                       (lah-dree-joh seh-RAH-mee-koh)

Timber   **Maderos**
                       (mah-DEH-rohs)

                       Madera de construcción
                       (mah-DEH-rah deh kohn-strook-SYOHN)

Tin   **Lata**
                       (LAH-tah)

	Chapa *(CHAH-pah)*
	Estaño *(ehs-TAH-nyoh)*
Toeboard	**Tabla de pie** *(TAH-blah deh PYEH)*
Toenail	**Clavo oblicuo** *(KLAH-voh oh-BLEE-koo-oh)*
Toilet	**Inodoro** *(ee-noh-DOH-roh)*
	Sanitario *(sah-nee-TAH-ryoh)*
	Escusado *(ehs-koo-SAH-doh)*
	Retrete *(reh-TREH-teh)*
Tongue and groove	**Machihembrado** *(mah-chee-ehm-BRAH-doh)*
Tools	**Herramientas** *(eh-rrah-MYEHN-tahs)*
Top-mount hangers	**Percha de montaje superior** *(PEHR-chah deh mon-TAH-heh soo-peh-RYOHR)*
Torch	**Antorcha** *(ahn-TOHR-chah)*
Townhouses	**Casas contiguas** *(KAH-sahs kohn-TEE-gwahs)*
Tracer	**Alambre testigo** *(ah-LAHM-breh teh-stee-goh)*
	Alambre rastreador *(ah-LAHM-breh Rahstreh-ah-dohr)*
Trap	**Sifón** *(see-FOHN)*
	Trampa hidráulica *(TRAHM-pah hee-DRAH-oo-lee-kah)*
Trap seal	**Sello de trampa hidráulica** *(SEH-joh deh TRAHM-pah hee-DRAH-oo-lee-kah)*
Travel distance	**Distancia de desplazamiento/ recorrido** *(dees-TAHN-syah deh dehs-plah-sah-MYEN-toh/rreh-koh-RREE-doh)*

Tread (stair)
    Huella
    (WEH-jah)
    Peldaño
    (pehl-DAH-nyoh)

Treated wood
    Madera tratada
    (mah-DEH-rah trah-TAH-dah)

Trench
    Zanja
    (SAHN-hah)

Trim
    Molduras
    (mohl-DOO-rahs)

Truss
    Cercha
    (SEHR-chah)
    Reticulado
    (reh-tee-koo-LAH-doh)
    Armadura
    (ar-mah-DOO-rah)
    Cabreada
    (kah-breh-AH-dah)
    Caballete
    (kah-bah-JEH-teh)

Tubing
    Cañería
    (kah-nyeh-REE-ah)
    Tubería
    (too-beh-REE-ah)

Tufted
    De bucles
    (deh BOO-klehs)

Turned-down footings
    Zapatas invertidas
    (sah-PAH-tahs een-vehr-TEE-dahs)

Unbalanced fill

Relleno sin consolidar
(reh-JEH-noh seen kohn-soh-lee-DAR)

Unbalanced loads

Cargas no balanceadas
(KAHR-gahs noh bah-lahn-seh-AH-dahs)

Uncased concrete piles

Pilotes sin encamisar/ sin camisa
(pee-LOH-tehs seen ehn-KAH-mee-SAR/ seen kah-MEE-sah)

Undercut

Resquicio
(rehs-KEE-syoh)

Underground

Subterráneo
(soob-teh-RRAH-neh-oh)

Underlap

Subsolape
(soob-soh-LAH-peh)

Underlayment

Capa base
(KAH-pah BAH-seh)

Capa de soporte
(KAH-pah deh soh-POHR-teh)

Capa bituminosa debajo del piso de madera
(KAH-pah bee-too-mee-NOH-sah deh-BAH-joh dehl PEE-soh deh mah-DEH-rah)

Substrato
(soob-STRAH-toh)

Bajopiso
(bah-joh-PEE-soh)

Union

Unión
(oon-YOHN)

Junta
(HOON-tah)

Unlimited area

Área ilimitada
(AH-reh-ah ee-lee-mee-TAH-dah)

Unsafe buildings	**Edificaciones inseguras** *(eh-dee-fee-kah-SYOH-nehs een-seh-GOO-rahs)*
Unstable ground	**Terreno inestable** *(teh-<u>RR</u>EH-noh ee-neh-STAH-bleh)*
Uplift (wind)	**Remonte** *(reh-MOHN-teh)* **Levantamiento (por viento)** *(leh-vahn-tah-MYEHN-toh (pohr VYEHN-toh)*
Urinal	**Urinario** *(oo-ree-NAH-ree-oh)* **Urinal** *(oo-ree-NAHL)* **Mingitorio** *(meen-hee-TOH-ree-oh)*
Use	**Uso** *(OO-soh)* **Utilizar** *(oo-tee-lee-SAR)*
Utility	**Utilidad** *(oo-tee-lee-DAHD)* **Uso general** *(OO-soh heh-neh-RAHL)*
Utility (public service)	**Servicios públicos** *(sehr-VEE-syohs POOH-blee-kohs)*

Vacuum	**Vacío** *(vah-SEE-oh)* **Aspiradora** *(ahs-pee-rah-DOH-rah)*
Vacuum breaker	**Interruptor de vacío** *(een-teh-rroop-TOHR deh vah-SEE-oh)*
Valuation	**Valuación** *(vah-luah-SYOHN)*
Value	**Valor** *(vah-LOHR)*
Valve, bleeder	**Válvula de purga** *(VAHL-voo-lah deh POOR-gah)*
Valve, hub	**Válvula de cubo** *(VAHL-voo-lah deh KOO-boh)*
Valve, key	**Válvula de llave** *(VAHL-voo-lah deh JAH-veh)*
Veneer	**Revestimiento** *(reh-veh-stee-MYEHN-toh)*
Veneer plaster	**Revestimiento de enlucido/revoque** *(reh-veh-stee-MYEHN-toh deh ehn-loo-SEE-doh/reh-VOH-keh)*
Vent	**Respiradero** *(reh-spee-rah-DEH-roh)*
Vent (v)	**Ventilar** *(vehn-tee-LAR)* **Evacuar** *(eh-vah-KWAR)* **Aliviar** *(ah-lee-VYAR)* **Desahogar** *(dehs-ah-oh-GAR)*

Vent pipe	**Tubo de ventilación**
	(TOO-boh deh vehn-tee-lah-SYOHN)
Vent shaft	**Recinto de ventilación**
	(reh-SEEN-toh deh vehn-tee-lah-SYOHN)
Vent stack	**Respiradero vertical**
	(reh-spee-rah-DEH-roh vehr-tee-KAHL)
Vent system	**Sistema de ventilación**
	(sees-TEH-mah deh vehn-tee-lah-SYOHN)
Ventilate	**Ventilar**
	(vehn-tee-LAR)
Venting system	**Sistema de evacuación**
	(sees-TEH-mah deh eh-vah-kwah-SYOHN)
Vertical pipe	**Tubería vertical**
	(too-beh-REE-ah vehr-tee-KAHL)
Vessel	**Recipiente**
	(reh-see-PYEHN-teh)
Vestibule	**Vestíbulo**
	(vehs-TEE-boo-loh)
Vinyl siding	**Revestimiento vinilo**
	(reh-veh-stee-MYEHN-toh vee-NEE-loh)
Vise	**Morsa**
	(MOHR-sah)
Void space	**Espacio vacío**
	(ehs-PAH-syoh vah-SEE-oh)
Volatile memory	**Memoria volátil**
	(meh-MOH-ree-ah voh-LAH-teel)
Voltage	**Voltaje**
	(vohl-TAH-heh)
Volts	**Voltios**
	(VOHL-tee-ohs)

Wafer head
Cabeza plana
(kah-BEH-sah PLAH-nah)

Wainscot, Wainscoting
Friso
(FREE-soh)

Alfarje
(ahl-FAR-heh)

Material de revestimiento
(mah-teh-RYAHL deh reh-veh-stee-MYEHN-toh)

Zócalos altos
(SOH-kah-lohs AHL-tohs)

Waive
Descartar
(dehs-kahr-TAHR)

Renunciar a un derecho
(reh-noon-SYAR- ah oon deh-REH-choh)

Walk-in cooler
Frigorífico
(free-goh-REE-fee-koh)

Walking surface
Superficie/Área peatonal
(soo-pehr-FEE-syeh/AH-reh-ah peh-ah-toh-NAHL)

Walks, moving
Caminos móviles
(kah-MEE-nohs MOH-vee-lehs)

Walkway
Camino
(kah-MEE-noh)

Wall
Muro
(MOO-roh)

Pared
(pa-REHD)

Barda
(BAR-dah)

Wall face
Placa de pared
(PLAH-cah deh pa-REHD)

Wall frames   **Estructuras de muros**
(ehs-trook-TOO-rahs deh MOO-rohs)

Wallboard   **Plancha de yeso**
(PLAHN-chah deh YEH-soh)

Cartón de yeso
(kahr-TOHN deh YEH-soh)

Warehouse   **Depósito**
(deh-POH-see-toh)

Bodega
(boh-DEH-gah)

Almacén
(ahl-mah-SEHN)

Galpón
(gahl-POHN)

Washer and Dryer   **Lavadora y secadora**
(lah-vah-DOH-rah ee seh-kah-DOH-rah)

Washer   **Arandela**
(ah-rahn-DEH-lah)

Planchuela de perno
(plahn-CHWEH-lah deh PEHR-noh)

Water heater   **Calentador/Calentón de agua**
(kah-leh-tah-DOHR/kah-lehn-TOHN de AH-gwah)

Calefón
(kah-leh-FOHN)

Termotanque
(tehr-moh-TAHN-keh)

Water main   **Tubería principal**
(too-beh-REE-ah preen-see-PAL)

Tubería matriz
(too-beh-REE-ah mah-TREES)

Water well   **Aljibe**
(ahl-HEE-beh)

Pozo de agua
(POH-soh deh AH-gwah)

Waste   **Desperdicios**
(dehs-pehr-DEE-syohs)

Wax seal   **Empaque de cera**
(ehm-PAH-keh deh SEH-rah)

Web   **Alma**
(AHL-mah)

	Membrana *(mehm-BRAH-nah)*
Web stiffener	**Atiesador de alma** *(ah-tyeh-sah-DOHR deh AHL-mah)*
Weeds	**Yerbajos** *(jehr-BAH-hos)*
Wedge	**Cuña** *(KOO-nyah)*
Welding	**Soldadura** *(sohl-dah-DOO-rah)*
Welding rod	**Electrodo** *(eh-lehk-TROH-doh)*
Well, dug	**Pozo excavado** *(POH-soh ehk-SKAH-vah-doh)*
Window	**Ventana** *(ven-TAH-nah)*
Window sill	**Soporte de ventana** *(so-POR-teh deh ven-TAH-nah)*
	Repisa de ventana *(reh-PEE-sah deh ven-TAH-nah)* *(see-mehn-ta-SYOHN)*
Wire	**Alambre** *(ah-LAM-breh)*
Wire backing	**Alambre de soporte** *(ah-LAM-breh deh soh-pohr-TEH)*
Wire, chicken	**Alambre de pollo** *(ah-LAM-breh deh POH-joh)*
Wire connectors	**Conectores de alambre** *(koh-nek-TOH-rehs deh ah-LAM-breh)*
	Cable alambre conector *(KAH-bleh ah-LAM-breh koh-nek-TOHR)*
Wire fabric	**Malla de alambre** *(MAH-jah deh ah-LAM-breh)*
Wire mesh	**Tela metálica** *(TEH-lah meh-TAH-lee-kah)*
Wire tie	**Alambre de paca** *(ah-LAM-breh deh PAH-kah)*
Wood board	**Tabla de madera** *(TAH-blah deh mah-DEH-rah)*

Wood framing	**Estructura en/de madera** *(eh-strook-TOO-rah ehn/deh mah-DEH-rah)* **Entramado de madera** *(ehn-trah-MAH-doh deh mah-DEH-rah)* **Bastidores de madera** *(bah-stee-DOH-rehs deh mah-DEH-rah)*
Wood members	**Elementos de madera** *(eh-leh-MEHN-tohs deh mah-DEH-rah)*
Wood shakes	**Duela de madera** *(DWEH-lah deh mah-DEH-rah)*
Wood shingles	**Tejas de madera** *(TEH-hahs deh mah-DEH-rah)*
Wood strip flooring	**Piso enlistonado de madera** *(PEE-soh ehn-lee-stoh-NAH-doh deh mah-DEH-rah)*
Wood truss	**Cercha de madera** *(SEHR-chah deh mah-DEH-rah)*
Woodworker	**Ebanista** *(eh-bah-NEES-tah)*
Work	**Obra** *(OH-brah)* **Trabajo** *(trah-BAH-hoh)*
Work, completion of	**Terminación de obra** *(tehr-mee-nah-SYOHN deh OH-brah)*
Working drawings	**Planos** *(PLAH-nohs)* **Dibujos** *(dee-BOO-hos)*

Electrical Terms

Accessible	**Accesible** *(ahk-seh-SEE-bleh)*
Adapter	**Adaptador** *(ah-dap-dah-DOHR)*
Amperes	**Amperios** *(ahm-PEH-ree-ohs)*
Amps	**Amperios** *(ahm-PEH-ree-ohs)*
Appliance	**Electrodoméstico** *(eh-lehk-troh-doh-MEHS-tee-kohs)* **Aparato eléctrico** *(ah-pah-RAH-toh eh-LEHK-tree-koh)*
Appliance circuit	**Circuito para aparatos** *(seer-KWEE-toh pah-rah ah-pah-RAH-tohs)*
Bare conductor	**Conductor desnudo** *(kohn-dook-TOHR dehs-NOO-doh)*
Bending machine, single operation	**Máquina de curvar de un solo golpe** *(MAH-kee-nah deh koor-VAHR deh oon SOH-loh GOHL-peh)*
Bend radius	**Radio de curvatura** *(RAH-dee-oh deh koor-vah-TOO-rah)*
Bonding jumper	**Puente de conexión equipotencial** *(PWEN-teh deh koh-nehk-SYOHN eh-kee-poh-ten-see-AL)* **Puente eléctrico** *(PWEN-teh eh-LEHK-tree-koh)*
Branch	**Rama** *(RAH-mah)*
Branch circuit	**Circuito secundario** *(seer-KWEE-toh seh-koon-DAH-ree-oh)* **Circuito derivado** *(seer-KWEE-toh deh-ree-VAH-doh)* **Ramal** *(rrah-MAHL)*

Busbar	**Barra colectora** *(BAH-rrah koh-lehk-TOH-rah)*
Circuit	**Circuito** *(seer-KWEE-toh)*
Circuit breaker	**Interruptor automático** *(een-teh-rroop-TOHR ah-oo-toh-MAH-tee-koh)*
Clamp	**Abrazadera** *(ah-brah-sah-DEH-rah)* **Mordaza** *(mohr-DAH-sah)*
Conductors	**Conductores** *(kohn-dook-TOH-rehs)*
Continuity checker	**Rastreador de continuidad** *(rahs-treh-ah-DOHR deh kohn-tee-nwee-DAHD)*
Continuity meter	**Multímetro** *(mool-TEE-meh-troh)*
Controller	**Controlador** *(kohn-troh-lah-DOHR)*
Cross sectional area	**Área común** *(AH-reh-ah koh-MOON)*
Delta-Wye	**Delta-estrella** *(DEHL-tah ehs-TREH-yah)*
Derating	**Factor de corrección** *(fahk-TOHR deh koh-rrehk-SYOHN)*
Device box	**Caja de dispositivos** *(KAH-hah deh dees-poh-see-TEE-vohs)*
Dustproof	**A prueba de polvo** *(ah-PRWEH-vah deh POHL-voh)*
Duty	**Servicio** *(sehr-VEE-syoh)*
Fault current	**Corriente de falla** *(koh-RYEHN-teh deh FAH-yah)* **Corriente de falta** *(koh-RYEHN-teh deh FAHl-tah)* **Corriente de pérdida** *(koh-RYEHN-teh deh PEHR-dee-dah)*
Feeder circuit	**Circuito de suministro** *(seer-KWEE-toh deh soo-mee-NEES-troh)*

Feeder tap	**Toma de corriente de circuito de suministro** *(TOH-mah deh koh RRYEHN- teh de seer-KWEE-toh deh soo-mee-NEES-troh)*
Field bend	**Curva hecha en obra** *(koor-vah EH-chah ehn OH-brah)*
Fittings	**Herrajes** *(eh-RRAH-hehs)*
Fixtures	**Aparatos** *(ah-pah-RAH-tohs)*
Galvanic action	**Reacción galvánica** *(reh-ahk-SYOHN gahl-VAH-nee-kah)*
Grounding	**De puesta a tierra** *(deh PWEHS-tah ah TYEH-rrah)*
Grounded	**Puesto a tierra** *(PWEHS-toh ah TYEH-rrah)*
Ground fault circuit	**Interruptor fusible de seguridad a tierra** *(een-teh-rroop-TOHR foo-SEE-bleh deh seh-goo-ree-DAHD ah TYEH-rra)*
Ground-fault circuit interrupter(GFCI)	**Interruptor contra faltas a tierra** *(een-teh-rroop-TOHR kohn-trah FAHL-tahs ah TYEH-rrah)*
Ground/neutral bus bar	**Bandeja neutra/a tierra** *(bahn-DEH-hah NEH-oo-trah/ah TYEH-rra)*
Guys	**Cuerdas** *(KWEHR-dahs)*
Hickey	**Adaptador** *(ah-dap-dah-DOHR)*
High-leg conductor	**Conductor de extremo alto** *(kohn-dook-TOHR deh ehks-TREH-moh AHL-toh)*
Inductive heat	**Calentamiento por inducción** *(kah-lehn-tah-MYEHN-toh pohr een-dook-SYOHN)*
Input voltage	**Tensión del suministro** *(tehn-SYOHN dehl soo-mee-NEES-troh)*

Inverse time circuit breaker	**Interruptor automático de retardo inverso** *(een-teh-rroop-TOHR ah-oo-toh-MAH-tee-koh deh reh-tahr-doh een-vehr-soh)*
Isolated	**Aislado** *(ah-ees-LAH-doh)*
Junction box	**Caja de bornes** *(KAH-hah deh BOHR-nehs)*
	Caja de conexiones de empalme *(KAH-hah deh koh-nek-SYOH-nehs deh ehm-PAHL-meh)*
KVA	**kVA** *(kah-veh-ah)*
	Kilo voltioamperios *(KEE-loh VOHL-tee-oh ahm-PEH-ree-ohs)*
	Kilovatios *(KEE-loh VAH-tee-ohs)*
	Kilowatts *(kee-loh-WATS)*
Lighting	**Iluminación** *(ee-loo-mee-nah-SYOHN)*
Live parts	**Partes en tensión** *(PAHR-tehs ehn tehn-SYOHN*
Lug	**Talón** *(tah-LOHN)*
	Asiento *(ah-SYEHN-toh)*
Made electrode	**Electrodo fabricado** *(eh-lek-TROH-doh fah-bree-KAH-doh)*
Molded case switch	**Interruptor de maniobra en caja moldeada** *(een-teh-rroop-TOHR deh mah-NYOH-brah ehn KAH-hah mohl-deh-AH-dah)*
	Interruptor de sección en caja moldeada *(een-teh-rroop-TOHR deh sehk-SYOHN ehn KAH-hah mohl-deh-AH-dah)*
Non-metallic sheathed cable	**Cable de cubierta no metálica** *(KAH-bleh deh koo-bee-EHR-tah noh meh-TAH-lee-kah)*

Nontime delay fuse

Fusible sin retardo
(foo-SEE-bleh seen reh-TAHR-doh)

**Objectionable
current flow**

Flujo de corriente inaceptable
*(FLOO-hoh de koh-ree-EHN-teh een-ah-
sehp-TAH-bleh)*

Open conductor

Conductores a la vista
(kohn-dook-TOH-rehs ah lah VEES-tah)

Outlet

Toma de corriente
(TOH-mah deh koh RRYEHN- teh)

Overcurrent

Sobreintensidad
(soh-breh een-tehn-see-DAHD)

**Overhead service
conductors**

**Conductores aéreos
de acometida**
*(kohn-dook-TOH-rehs ah-EH-reh-ohs deh
ah-koh-meh-TEE-dah)*

Overhead spans

Tramos aéreos
(TRAH-mohs ah-EH-reh-ohs)

Overload

Sobrecarga
(soh-breh-KAHR-gah)

Panelboard

Panel de distribución
(pah-NEHL deh dees-tree-boo-SYOHN

Tablero de distribución
(tah-BLEH-roh deh dees-tree-boo-SYOHN

Cuadro general
(KWAH-droh heh-neh-RAHL)

Pressure connectors

Conectores a presión
(koh-nehk-TOH-rehs ah preh-SYOHN)

Push box

Pulsador
(pool-sah-DOHR)

Raceway

Canalización
(kah-nah-lee-sah-SYOHN)

Rainproof

A prueba de lluvia
(ah PRWEH-bah deh JOO-vee-ah)

Resistente a la lluvia
(reh-sees-TEHN-teh ah lah JOO-vee-ah)

Rated

Nominal
(noh-mee-NAHL)

Rating	**Régimen nominal** *(REH-hee-mehn noh-mee-NAHL)*
	Potencia de servicio *(poh-TEHN-syah deh sehr-VEE-syoh)*
Readily accessible	**Fácilmente accesible** *(FAH-seel-mehn-teh ahk-seh-SEE-bleh)*
Reamed	**Abocardado** *(ah-boh-kahr-DAH-doh)*
	Escariado *(ehs-kah-ree-AH-doh)*
Receptacle	**Base de toma de corriente** *(BAH-seh deh TOH-mah deh koh <u>RR</u>YEHN-teh)*
	Toma de corriente *(TOH-mah deh koh <u>RR</u>YEHN- teh)*
	Enchufe *(ehn-CHOO-feh)*
RMC	**Tubos de metal rígidos** *(TOO-bohs deh meh-TAHL <u>RR</u>EE-hee-dohs)*
Schedule	**Plan** *(plahn)*
	Cronograma *(kroh-noh-GRAH-mah)*
Service	**Acometida** *(ah-koh-meh-TEE-dah)*
Service drop	**Acometida aérea** *(ah-koh-meh-TEE-dah ah-EH-reh-ah)*
Service conductors	**Conductores de la acometida** *(kohn-dook-TOH-rehs deh lah ah-koh-meh-TEE-dah)*
Service-entrance	**Entrada a la acometida** *(ehn-TRAH-dah deh lah ah-koh-meh-TEE-dah)*
Service disconnect	**Desconexión de la acometida** *(dehs-koh-nehk-SYOHN deh lah ah-koh-meh-TEE-dah)*
Service lateral	**Acometida subterránea** *(ah-koh-meh-TEE-dah soob-teh-RRAH-neh-ah)*
Service mast	**Poste de acometida** *(POHS-teh de ah-koh-meh-TEE-dah)*

Shunt field   **Campo inductor en derivación**
(KAHM-poh een-dook-TOHR ehn deh-ree-bah-SYOHN)

Single phase   **Monofásica**
(moh-noh-FAH-see-kah)

Size (conductors)   **Sección**
(sehk-SYOHN)

**Smallest size
(conductors)**   **Sección inferior**
(sehk-SYOHN een-feh-ree-OHR)

Snap switches   **Interruptores de acción rápida**
(een-teh-rroop-TOHR-ehs deh ahk-SYOHN RAH-pee-dah)

Squirrel-cage rotor   **Rotor de cortocircuito**
(roh-TOHR deh kohr-toh-seer-KWEE-toh)

            Rotor de jaula de ardilla
(roh-TOHR deh ha-oo-lah deh ahr-DEE-jah)

**Step down
voltage transformer**   **Transformador reductor
de voltaje**
(trahns-fohr-mah-DOHR reh-dook-TOHR deh vohl-TAH-heh)

Switch   **Interruptor**
(een-teh-rroop-TOHR)

Switchboard   **Cuadro de distribución**
(KWAH-droh deh dees-tree-boo-SYOHN

Switch leg   **Rama local**
(RAH-mah loh-KAHL)

Tap   **Toma de corriente**
(TOH-mah deh koh RRYEHN- teh)

Three phase   **Trifásico**
(tree-FAH-see-koh)

Three (3)-wire   **Tripolar**
(tree-poh-LAHR)

Torque motor   **Motor de baja velocidad**
(moh-TOHR deh BAH-hah veh-loh-see-DAHD)

Underground   **Subterráneos**
(soob-teh-RRAH-neh-ohs)

Ungrounded	**No puesto a tierra** *(NOH PWEHS-toh ah TYEH-rrah)*
	Sin puesta a tierra *(seen PWEHS-tah ah TYEH-rrah)*
Volt amps (VA)	**Vatios** *(VAH-tee-ohs)*
	Voltioamperios *(vohl-tee-oh-ahm-PEH-ree-ohs)*
Volts	**Voltios** *(VOHL-tee-ohs)*
Waterproof	**A prueba de agua** *(ah PRWEH-bah deh AH-gwah)*
	Resistente al agua *(reh-sees-TEHN-teh ahl AH-gwah)*
Wire	**Alambre** *(ah-LAHM-breh)*
	Cable *(KAH-bleh)*
Wiring	**Cables** *(KAH-blehs)*
Wiring method	**Método de instalación** *(MEH-toh-doh deh eens-tah-lah-SYOHN)*

Mechanical Terms

Abrasive materials
    Materiales abrasivos
    (Mah-teh-RYALH-lehs ah-brah-SEE-vos)

Absorption
    Absorción
    (Ahb-sohr-SYOHN)

Air
    Aire
    (AH-ee-reh)

Air conditioning
    Aire acondicionado
    (AH-ee-reh)

Air distribution system
    Sistema de distribución de aire
    (sees-TEH-mah deh dees-tree-boo-SYOHN deh AH-ee-reh)

Air makeup
    Aire de reposición
    (AH-ee-reh deh reh-poh-see-SYOHN)

Air-space clearance
    Espacio de aire libre
    (ehs-PAH-syoh deh AH-ee-reh LEE-breh)

Air-handling
    Manejo de aire
    (mah-NEH-hoh deh AH-ee-reh)

Alteration
    Modificación
    (moh-dee-fee-kah-SYOHN)

Ampacity
    Intensidad máxima admisible
    (een-tehn-see-DAHD MAHK-see-mah ahd-mee-SEE-bleh)

Appliance
    Aparato
    (ah-pah-RAH-toh)
    Artefacto
    (ahr-teh-FAHK-toh)

Appliance type
    Tipo de artefacto
    (deh ahr-teh-FAHK-toh)

Appliance, existing
    Artefacto existente
    (ahr-teh-FAHK-toh ehk-sees-TEHN-teh)

Appliance, vented
    Artefacto con ventilación
    (ahr-teh-FAHK-toh kohn vehn-tee-lah-SYOHN)

Automatic boiler
    Caldera automática
    (kahl-DEH-rah ah-oo-toh-MAH-tee-kah)

Backdraft
Contratiro
(kohn-trah-TEE-roh)

Backdraft damper
Regulador de tiro de contratiro
(reh-goo-lah-DOHR deh TEE-roh deh kohn-trah-TEE-roh)

Backshelf hood
Campana trasera
(kahm-PAH-nah trah-SEH-rah)

Baseboard convector
Convector de zócalo
(kohn-vehk-TOHR deh SOH-kah-loh)

Booster fan
Ventilador reforzador
(vehn-tee-lah-DOHR reh-fohr-sah-DOHR)

Boot
Accesorio de transición
(ahk-seh-SOH-ryoh deh trahn-see-SYOHN)

Brazed, joint
Junta soldada en fuerte
(HOON-tah sohl-DAH-dah ehn FWER-teh)

Brazing
Soldadura en fuerte
(sohl-dah-DOO-rah ehn FWER-teh)

Btu
Btu (Unidad Térmica Británica)
(beh-teh-oo [oo-nee-DAHD TEHR-mee-kah bree-TAH-nee-kah])

Burner
Quemador
(keh-mah-DOHR)

Bypass valve
Válvula de derivación
(VAHL-voo-lah deh deh-ree-vah-SYOHN)

Casing
Carcasa
(kahr-KAH-sah)

Chimney connector
Conector de chimenea
(koh-nehk-TOHR deh chee-meh-NEH-ah)

Chimney passageway
Pasillo de chimenea
(pah-SEE-joh deh chee-meh-NEH-ah)

Closed combustion
Combustión cerrada
(kohm-boos-TYOHN seh-RRAH-dah)

Clothes dryer
Secadora de ropas
(seh-kah-DOH-rah deh RROH-pahs)

Combustible assembly
Conjunto combustible
(kohn-HOON-toh kohm-boos-TEE-bleh)

Combustible liquids
Líquidos combustibles
(LEE-kee-dohs kohm-boos-TEE-blehs)

Combustible material   **Material combustible**
(mah-teh-RYHAL kohm-boos-TEE-bleh)

Combustion   **Combustión**
(kohm-boos-TYOHN)

Combustion air   **Aire de combustión**
(AH-ee-reh deh kohm-boos-TYOHN)

Combustion chamber   **Cámara de combustión**
(KAH-mah-rah deh kohm-boos-TYOHN)

Combustion products   **Productos de combustión**
(proh-DOOK-tohs deh kohm-boos-TYOHN)

**Commercial cooking
appliances**   **Artefactos de cocina comercial**
(ahr-teh-FAHK-tohs deh koh-SEE-nah koh-mehr-SYAHL)

Concealed location   **Localización oculta**
(loh-kah-lee-sah-SYOHN oh-KOOL-tah)

Condensate   **Condensados**
(kohn-den-SAH-dohs)

Condenser   **Condensador**
(kohn-den-sah-DOHR)

Condensing unit   **Unidad condensadora**
(oo-nee-DAHD kohn-den-sah-DOH-rah)

Conditioned space   **Espacio acondicionado**
(ehs-PAH-syoh ah-kohn-dee-syoh-NAH-doh)

Confined space   **Espacio confinado**
(ehs-PAH-syoh kohn-fee-NAH-doh)

Contact lap   **Solapadura de contacto**
(soh-lah-pah-DOO-rah deh kohn-TAHK-toh)

Control   **Control**
(kohn-ROHL)

Conversion burner   **Quemador de conversión**
(keh-mah-DOHR deh kohn-vehr-SYOHN)

Cooling equipment   **Equipo de enfriamiento**
(eh-KEE-poh deh ehn-FRYAH-myehn-toh)

Cooking appliance   **Artefacto de cocina**
(ahr-teh-FAHK-toh deh koh-SEE-nah)

Crawl space   **Espacio angosto**
(ehs-PAH-syoh ahn-GOHS-toh)

Cutoff	**Cierre** *(SYEH-rreh)*
Direct-vent appliance	**Artefactos de ventilación directa** *(ahr-teh-FAHK-toh deh vehn-tee-lah-SYOHN dee-REHK-tah)*
Double island canopy hood	**Campana de doble isla** *(kahm-PAH-nah deh DOH-bleh EES-lah)*
Down draft	**Tiro descendente** *(TEE-roh deh-sehn-DEN-teh)*
Draft	**Tiro** *(TEE-roh)*
Drilling	**Perforación** *(pehr-foh-rah-SYOHN)*
Drip	**Colector** *(koh-lehk-TOHR)*
Dry cleaning	**Limpieza en seco** *(leem-PYEH-sah ehn SEH-koh)*
Duct heater	**Calefactor de conducto** *(kah-leh-fahk-TOHR deh kohn-DOOK-toh)*
Duct system	**Sistema de conductos** *(sees-TEH-mah deh kohn-DOOK-tohs)*
Evaporative cooler	**Enfriador evaporativo** *(ehn-fryah-DOHR eh-vah-poh-rah-TEE-voh)*
Evaporative cooling	**Enfriamiento evaporativo** *(ehn-fryah-MYEHN-toh eh-vah-poh-rah-TEE-voh)*
Evaporator	**Evaporador** *(eh-vah-poh-rah-DOHR)*
Excess air	**Aire en exceso** *(AH-ee-reh ehn ehk-SEH-soh)*
Exfiltration	**Exfiltración** *(ehks-feel-trah-SYOHN)*
Exhaust air	**Aire de extracción** *(AH-ee-reh deh ehks-trahk-SYOHN)*
Exhaust rate	**Gasto de extracción** *(GAHS-toh deh ehks-trahk-SYOHN)*

Exhaust system	**Sistema de extracción** *(sees-TEH-mah deh ehks-trahk-SYOHN)*
Exhauster	**Aspirador** *(ahs-pee-rah-DOHR)*
Extra-heavy duty	**De uso muy pesado** *(deh OO-soh MOO-ee peh-SAH-doh)*
Eyebrow hood	**Campana de tipo ceja** *(kahm-PAH-nah deh TEE-poh SEH-ha)*
Factory-built fireplace	**Chimenea prefabricada** *(chee-meh-NEH-ah preh-fah-bree-KAH-dah)* **Hogar prefabricado** *(oh-GAR preh-fah-bree-KAH-doh)*
Fan	**Ventilador** *(vehn-tee-lah-DOHR)*
Fireplace	**Hogar** *(oh-GAHR)*
Fireplace stove	**Hogar tipo estufa** *(oh-GAHR TEE-poh ehs-TOO-fah)*
Flame safeguard	**Salvaguarda de llama** *(sal-vah-GWAHR-dah deh YAH-mah)*
Flame spread index	**Índice de propagación de llama** *(EEN-dee-seh deh proh-pah-gah-SYOHN deh YAH-mah)*
Flanged fitting	**Accesorio embridado** *(ahk-seh-SOH-ryoh em-bree-DAH-doh)*
Flash	**Babeta** *(bah-BEH-tah)*
Flash point	**Punto de ignición** *(POON-toh deh ee-gnee-SYOHN)*
Floor area, net	**Área neta de piso** *(AH-reh-ah NEH-tah deh PEE-soh)*
Flue liner	**Revestimiento de conducto de humo** *(reh-vehs-tee-MYEHN-toh deh kohn-DOOK-toh deh OO-moh)* **Revestimiento interior** *(reh-vehs-tee-MYEHN-toh een-teh-RYOHR)*
Forced air	**Aire forzado** *(AH-ee-reh forh-SAH-doh)*

Forced-draft venting   **Ventilación de tiro forzado**
(vehn-tee-lah-SYOHN deh TEE-roh fohr-SAH-doh)

Fuel gas   **Gas combustible**
(gahs kohm-boos-TEE-bleh)

Fuel oil   **Petróleo**
(peh-TROH-leh-oh)

Furnace   **Calefactor**
(kah-leh-fahk-TOHR)

Furnace blower   **Ventilador del calefactor**
(vehn-tee-lah-DOHR deh kah-leh-fahk-TOHR)

Furnace room   **Cuarto del calefactor**
(KWAHR-to dehl kah-leh-fahk-TOHR)

Glass gauge   **Tubo indicador**
(TOO-boh een-dee-kah-DOHR)

**Ground source
heat pump**   **Bomba de calor geotérmica**
(BOHM-bah deh kah-LOHR heh-oh-TEHR-mee-kah)

Hazardous location   **Localización peligrosa**
(loh-kah-lee-sah-SYOHN peh-lee-GROH-sah)

Hearth   **Hogar**
(oh-GAHR)

Heat exchanger   **Intercambiador de calor**
(een-tehr-kahm-byah-DOHR deh kah-LOHR)

Heat pump   **Bomba de calor**
(BOHM-bah deh kah-LOHR)

Heat transfer liquid   **Líquido de transferencia
de calor**
(LEE-kee-doh deh trahns-feh-REHN-syah deh kah-LOHR)

Heavy duty   **De uso pesado**
(deh OO-soh peh-SAH-doh)

High-heat appliance   **Artefacto de alto calor**
(ahr-teh-FAHK-toh deh AHL-toh kah-LOHR)

Hood   **Campana**
(kahm-PAH-nah)

Ignition source	**Fuente de ignición** *(FWEHN-teh deh ee-gnee-SYOHN)*
Induced-draft venting	**Ventilación de tiro inducido** *(vehn-tee-lah-SYOHN deh TEE-roh een-doo-SEE-doh)*
Joint, flanged	**Junta bridada** *(HOON-tah bree-DAH-dah)*
Joint, flared	**Junta abocinada** *(HOON-tah ah-boh-see-NAH-dah)*
Joint, mechanical	**Junta, mecánica** *(HOON-tah meh-KAH-nee-kah)*
Joint, plastic adhesive	**Junta de adhesivo plástico** *(HOON-tah deh ah-deh-SEE-voh PLAHS-tee-koh)*
Joint, soldered	**Junta soldada** *(HOON-tah sohl-DAH-dah)*
Joint, welded	**Junta soldada por fusión** *(HOON-tah sohl-DAH-dah)*
Label	**Sello** *(SEH-joh)*
Labeled	**Sellado** *(seh-JAH-doh)*
Light-duty	**De uso ligero** *(deh OO-soh lee-HEH-roh)*
Limit control	**Control de límites** *(kohn-TROHL deh LEE-mee-tehs)*
Limited charge system	**Sistema de carga limitada** *(sees-TEH-mah deh KAHR-gah lee-mee-TAH-dah)*
Lining	**Revestimiento interior** *(reh-vehs-tee-MYEHN-toh een-teh-RYOHR)*
Living space	**Espacio habitado** *(ehs-PAH-syoh ah-bee-TAH-doh)*
Load bearing member	**Miembro de carga** *(MYEHM-broh deh KAHR-gah)*
Location	**Ubicación** *(oo-beek-ah-SYOHN)*

Lower explosive limit (LEL)
Límite explosivo inferior
(LEE-mee-teh ehks-ploh-SEE-voh een-feh-RYOHR)

Lower flammability limit (LFL)
Límite inferior de inflamabilidad
(LEE-mee-teh een-feh-RYOHR deh een-flah-mah-bee-lee-DAHD)

Mechanical draft
Tiro mecánico
(sees-TEH-mah deh TEE-roh meh-KAH-nee-kph)

Mechanical exhaust system
Sistema de extracción mecánica
(sees-TEH-mah deh ehks-trahk-SYOHN meh-KAH-nee-kah)

Mechanical joint
Junta mecánica
(HOON-tah meh-KAH-nee-kah)

Mechanical system
Sistema mecánico
(sees-TEH-mah meh-KAH-nee-koh)

Medium-duty
De uso medio
(deh OO-soh MEH-dyoh)

Medium-heat appliance
Artefacto de mediano calor
(ahr-teh-FAHK-toh deh meh-DYAH-noh kah-LOHR)

Metal chimney
Chimenea de metal
(chee-meh-NEH-ah deh meh-TAHL)

Modular boiler
Caldera modular
(kahl-DEH-rah moh-doo-LAHR)

Natural draft
Tiro natural
(TEE-roh nah-too-RAHL)

Outdoor air
Aire exterior
(AH-ee-reh ehk-steh-RYOHR)

Outdoor opening
Abertura exterior
(ah-behr-TOO-rah ehk-steh-RYOHR)

Outlet
Boca de salida
(BOH-kah deh sah-LEE-dah)

Output
Rendimiento
(rehn-dee-MYEHN-toh)

Output rating	**Clasificación de rendimiento** *(klah-see-fee-kah-SYOHN deh rehn-dee-MYEHN-toh)*
Piping	**Tubería** *(too-beh-REE-ah)*
Plastic, thermoplastic	**Plástico termoplástico** *(PLAHS-tee-koh tehr-moh-PLAHS-tee-koh)*
Plastic, thermosetting	**Plástico termoendurecido** *(PLAHS-tee-koh tehr-moh-ehn-doo-reh-SEE-doh)*
Plenum	**Pleno** *(PLEH-noh)*
Power venting	**Ventilación mecánica** *(vehn-tee-lah-SYOHN meh-KAH-nee-kah)*
Pressure relief device	**Dispositivo de alivio de presión** *(dees-poh-see-TEE-voh deh ah-LEE-vyoh deh)*
Pressure relief valve	**Válvula de alivio de presión** *(VAHL-voo-lah deh - deh preh-SYOHN)*
Pressure vessels	**Recipientes a presión** *(reh-see-PYEHN-teh ah preh-SYOHN)*
Pressure-limiting device	**Dispositivo limitador de presión** *(dees-poh-see-TEE-voh lee-mee-tah-DOHR deh preh-SYOHN)*
Purge	**Purgar** *(poor-GAHR)*
Quick-opening valve	**Válvula de apertura rápida** *(VAHL-voo-lah deh ah-pehr-TOO-rah RRAH-pee-dah)*
Radiant heater	**Calefactor radiante** *(kah-leh-fahk-TOHR rah-DYAHN-teh)*
Radiant heating	**Calefacción radiante** *(kah-leh-fahk-SYOHN rah-DYAHN-teh)*
Range	**Cocina económica** *(koh-SEE-nah eh-koh-NOH-mee-kah)*
Rated capacity	**Capacidad nominal** *(kah-pah-see-DAHD noh-mee-NAHL)*
Rated output capacity	**Capacidad nominal de salida** *(kah-pah-see-DAHD noh-mee-NAHL deh sah-LEE-dah)*

Receiver, liquid	**Receptor de líquido** *(reh-sehp-TOHR deh LEE-kee-doh)*
Recirculated air	**Aire recirculado** *(AH-ee-reh reh-seer-koo-LAH-doh)*
Recirculating system	**Sistema de recirculación** *(sees-TEH-mah deh reh-seer-koo-lah-SYOHN)*
Reclaimed refrigerants	**Refrigerantes regenerados** *(reh-free-heh-RAHN-tehs reh-he-neh-RAH-dohs)*
Refrigerated room	**Cuarto refrigerado** *(KWAR-toh reh-free-heh-RAH-doh)*
Refrigerating system	**Sistema de refrigeración** *(sees-TEH-mah deh reh-free-heh-rah-SYOHN)*
Return air system	**Sistema de retorno de aire** *(sees-TEH-mah deh reh-TOHR-noh deh AH-ee-reh)*
Room heater vented	**Calefactor de cuarto con ventilación** *(kah-leh-fahk-TOHR deh KWAHR-toh kohn vehn-tee-lah-SYOHN)*
Saddle joint	**Junta saliente** *(HOON-tah sah-LYEHN-teh)*
Safeguard device	**Dispositivo de seguridad** *(dees-poh-see-TEE-voh deh seh-goo-ree-DAHD)*
Safety valve	**Válvula de seguridad** *(VAHL-voo-lah deh seh-goo-ree-DAHD)*
Self-contained	**Autocontenido** *(ah-oo-toh-kohn-teh-NEE-doh)*
Shaft enclosure	**Cerramiento de recinto** *(see-rrah-MYEHN-toh deh reh-SEEN-toh)*
Shield plate	**Placa de defensa** *(PLAH-kah deh deh-FEN-sah)*
Solid-fuel	**Combustibles sólidos** *(kohm-boos-TEE-blehs SOH-lee-dohs)*
Stacked washer	**Arandela entongada** *(ar-ahn-DEH-lah ehn-tohn-GAH-da)*

Stop valve **Válvula de cierre**
(VAHL-voo-lah deh SYEH-rreh)

Supply air **Aire de suministro**
(AH-ee-reh deh soo-mee-NEES-troh)

Supply plenum **Pleno de suministro**
(PLEH-noh deh soo-mee-NEES-troh)

Theoretical air **Aire teórico**
(AH-ee-reh teh-OH-ree-koh)

Thermal resistance (R) **Resistencia térmica (R)**
(reh-sees-TEHN-syah TEHR-mee-kah)

Tightfitting **Hermético**
(ehr-MEH-tee-koh)

Toxicity classification **Clasificación de toxicidad**
(klah-see-fee-kah-SYOHN deh tohk-see-see-DAHD)

Vent **Respiradero**
(rehs-pee-rah-DEH-roh)

Vent connector **Conector de respiradero**
(deh rehs-pee-rah-DEH-roh)

Vent damper **Regulador de tiro de respiradero**
(reh-goo-lah-DOHR deh TEE-roh deh rehs-pee-rah-DEH-roh)

Vented floor furnace **Calefactor de piso con ventilación**
(kah-leh-fahk-TOHR deh PEE-soh kohn vehn-tee-lah-SYOHN)

Ventilation **Ventilación**
(vehn-tee-lah-SYOHN)

Ventilation air **Aire de ventilación**
(AH-ee-reh deh vehn-tee-lah-SYOHN)

Ventilation rate **Gasto de ventilación**
(GAHS-toh deh vehn-tee-lah-SYOHN)

Venting system **Sistema de ventilación**
(sees-TEH-mah deh vehn-tee-lah-SYOHN)

Volume damper **Regulador de tiro de volumen**
(reh-goo-lah-DOHR deh TEE-roh deh voh-LOO-mehn)

111

Wall canopy hood

Campana de muro
(kahm-PAH-nah deh MOO-roh)
(kah-leh-fahk-TOHR deh AH-ee-reh deh)

Working space

Espacio de trabajo
(ehs-PAH-syoh deh trah-BAH-hoh)

**Working space
clearance**

Espacio libre de trabajo
*(ehs-PAH-syoh LEE-breh deh trah-BAH-
hoh)*

Plumbing Terms

Access cover
Tapa de acceso
(TAH-ph deh ahk-SEH-soh)

Air admittance valve
Válvula de admisión de aire
(VAHL-voo-lah deh ahd-mee-SYOHN deh AH-ee-reh)

Air break
Interruptor de aire
(een-teh-rroop-TOHR deh AH-ee-reh)

Airflow rate
Tasa de flujo de aire
(TAH-sah deh FLOO-hoh deh AH-ee-reh)

Tasa de gasto de aire
(TAH-sah deh deh AH-ee-reh)

Air gap
Espacio de aire
(ehs-PAH-syoh deh AH-ee-reh)

Area drain
Desagüe de área
(deh-SAH-gweh deh AH-reh-ah)

Desagüe de patio
(deh-SAH-gweh deh PAH-tee-oh)

Resumidero
(reh-soo-me-DEH-roh)

Aspirator
Aspirador
(ahs-pee-rah-DOHR)

Backflow
Contraflujo
(kohn-trah-FLOO-hoh)

Backflow connection
Conexión de contraflujo
(koh-nek-SYOHN deh kohn-trah-FLOO-hoh)

Backflow preventer
Válvula de contraflujo
(VAHL-voo-lah deh kohn-trah-FLOO-hoh)

Interruptor de contraflujo
(een-teh-rroop-TOHR deh kohn-trah-FLOO-hoh)

Backpressure, low head
Contraflujo, salto bajo
(kohn-trah-FLOO-hoh, SAHL-toh BAH-hoh)

Backsiphonage
Contrasifonaje
(kohn-trah-see-pho-NAH-heh)

Backwater valve
Válvula de contrapresión
(VAHL-voo-lah deh kohn-trah-preh-SYOHN)

Ball cock	**Llave de flotador** *(JAH-veh deh floh-tah-DOHR)*
Bracket	**Ménsula** *(MEHN-soo-lah)*
Branch	**Ramal** *(rah-MAHL)*
Bathroom group	**Grupo de muebles sanitarios** *(GROO-poh deh MWEH-blehs sah-nee-TAH-ryohs)*
Building drain	**Desagüe de la edificación** *(deh-SAH-gweh deh lah eh-dee-fee-kah-SYOHN)* **Resumidero** *(reh-soo-me-DEH-roh)*
Building subdrain	**Tubo de avenamiento de la edificación** *(TOO-boh ah-veh-nah-MYEN-toh deh lah eh-dee-fee-kah-SYOHN)*
Building trap	**Trampa hidráulica de la edificación** *(TRAHM-pah ee-DRAH-oo-lee-kah deh lah eh-dee-fee-kah-SYOHN)*
Branch vent	**Respiradero de ramal** *(rehs-pee-rah-DEH-roh deh rah-MAHL)*
Cesspool	**Sumidero** *(soo-me-DEH-roh)*
Circuit vent	**Respiradero en circuito** *(rehs-pee-rah-DEH-roh en seer-KWEE-toh)*
Cistern	**Cisterna** *(sees-TEHR-nah)*
Cleanout	**Registro** *(reh-HEES-troh)*
Coextruded	**Coextruido** *(koh-eks-troo-EE-doh)*
Combination fixture	**Artefacto de combinación** *(ahr-teh-FAHK-toh deh kohm-bee-nah-SYOHN)*
Combined drain	**Desagüe sanitario combinado** *(deh-SAH-gweh sah-nee-TAH-ryoh khom-bee-NAH-doh)*

Common vent   **Respiradero común**
                            (rehs-pee-rah-DEH-roh)

Critical level   **Nivel crítico**
                            (nee-VEHL KREE-tee-koh)

Culvert   **Alcantarilla**
                            (alkan-tah-REE-jah)
                            Desagüe
                            (deh-SAH-gweh)

Developed length   **Longitud desarrollada**
                            (lohn-hee-TOOD deh-sah-rroh-JAH-dah)

Discharge pipe   **Tubo de descarga**
                            (TOO-boh deh dehs-KAHR-gah)

Drain   **Desagüe**
                            (deh-SAH-gweh)

Drainage fitting   **Accesorio de desagüe sanitario**
                            *(ahk-seh-SOH-ryoh deh deh-SAH-gweh
                            sah-nee-TAH-ryoh)*

Drainage system   **Sistema de desagüe sanitario**
                            *(sees-TEH-mah deh deh-SAH-gweh sah-
                            nee-TAH-ryoh)*

Elastomeric gasket   **Empaque elastomérico**
                            (ehm-PAH-keh eh-lahs-toh-MEH-koh)

Expansion joint   **Junta de expansión**
                            (HOON-tah deh eks-pahn-SYOHN)

Faucet   **Llave**
                            (JAH-veh)

Fill valve   **Válvula de llenado**
                            (VAHL-voo-lah deh jeh-NAH-doh)
                            Válvula de alimentación
                            *(VAHL-voo-lah deh ah-lee-mehn-tah-
                            SYOHN)*

Fixture   **Artefacto**
                            (ahr-teh-FAHK-toh)

Fixture branch   **Ramal de artefactos**
                            (rah-MAHL deh ahr-teh-FAHK-tohs)

Fixture drain   **Desagüe de artefactos**
                            (deh-SAH-gweh deh ahr-teh-FAHK-toh)

Fixture fitting   **Accesorio de artefacto**
                            (ahk-seh-SOH-ryoh deh ahr-teh-FAHK-toh)

115

Flexible joint	**Junta flexible** *(HOON-tah flek-SEE-bleh)*
Fixture supply	**Alimentación de artefactos** *(ah-lee-mehn-tah-SYOHN deh ahr-teh-FAHK-tohs)*
Flood level rim	**Nivel de inundación** *(nee-VEHL deh ee-noon-dah-SYOHN)*
Flow pressure	**Presión de flujo** *(preh-SYOHN deh FLOO-hoh)*
Flush tank	**Tanque de inundación** *(TAHN-keh deh ee-noon-dah-SYOHN)*
Flush valve	**Válvula de baldeo** *(VAHL-voo-lah deh bahl-DEH-oh)*
Flushometer valve	**Válvula fluxómetro** *(VAHL-voo-lah deh flook-SOH-meh-tro)*
Flushometer tank	**Tanque fluxómetro** *(TAHN-keh flook-SOH-meh-tro)*
Forged	**Forjado** *(fohr-HAH-doh)*
Gate valve	**Válvula de compuerta** *(VAHL-voo-lah deh kohm-PWEHR-tah)*
Grease interceptor	**Interceptor de grasa** *(een-tehr-sehp-TOHR deh GRAH-sah)*
Grease trap	**Colector de grasas** *(koh-lehk-TOHR deh GRAH-sahs)*
Gutters	**Canaletas** *(kah-nah-LEH-tahs)*
Hydrants	**Hidrantes** *(ee-DRAHN-tehs)*
Horizontal branch drain	**Ramal sanitario horizontal** *(rah-MAHL sah-nee-TAH-ryoh oh-ree-sohn-TAHL)*
Hot water	**Agua caliente** *(AH-gwah kah-LYEHN-teh)*
House trap	**Trampa doméstica** *(TRAHM-pah doh-MEHS-tee-kah)*
Indirect waste pipe	**Tubo de desagüe indirecto** *(TOO-boh deh deh-SAH-gweh een-dee-REHK-toh)*

Individual vent

Respiradero individual
(rehs-pee-rah-DEH-roh een-dee-vee-DWAL)

Individual water supply

Abastecimiento de agua individual
(ah-bahs-teh-see-MYEHN-toh deh AH-gwah een-dee-vee-DWAL)

Leader

Tubo de bajada
(TOO-boh deh bah-ha-dah)

Leakage

Fugas
(FOO-gahs)

Macerating toilet systems

Sistemas sanitarios de macerado
(sees-TEH-mahs sah-nee-TAH-ryohs deh mah-seh-RAH-doh)

Manifold

Distribuidor
(dees-tree-boo-ee-DOHR)

Manifold systems

Sistemas múltiples de distribución
(sees-TEH-mahs MOOL-tee-plehs deh dees-tree-boo-SYOHN)

Mechanical joint

Junta mecánica
(HOON-tah meh-KAH-nee-kah)

Notching

Entalladura
(ehn-tah-jah DOO-rah)

Open air

Aire libre
(AH-ee-reh LEE-breh)

Plumbing

Instalaciones hidráulicas y sanitarias
(een-stah-lah-SYOH-nehs ee-DRAH-oo-lee-kahs ee sah-nee-TAH-ryahs)

Instalaciones hidrosanitarias
(een-stah-lah-SYOH-nehs ee-DROH-sah-nee-TAH-ryahs)

Plumbing appliance

Mueble sanitario
(MWEH-bleh sah-nee-TAH-ryoh)

Plumbing appurtenance

Accesorio sanitario
(ahk-seh-SOH-ryoh sah-nee-TAH-ryoh)

Plumbing fixture

Artefacto sanitario
(ahr-teh-FAHK-toh sah-nee-TAH-ryoh)

Potable water

Agua potable
(AH-gwah poh-TAH-bleh)

Public water main

Tubo matriz público
(TOO-boh mah-TREES POO-blee-koh)

Quick closing valve

Válvula de acción rápida
(VAHL-voo-lah deh ahk-SYOHN RRAH-pee-dah)

Rate of flow

Tasa de flujo
(TAH-sah deh FLOO-hoh)

Tasa de gasto
(TAH-sah deh GAHS-toh)

Reduced pressure

Presión reducida
(preh-SYOHN re-doo-SEE-dah)

Relief valve

Válvula de alivio
(VAHL-voo-lah deh)

Relief vent

Respiradero de alivio
(rehs-pee-rah-DEH-roh deh)

Rim

Borde
(BOHR-deh)

Riser

Tubería vertical montante
(too-beh-REE-ah vehr-tee-KAHL mohn-TAHN-teh)

Riser water pipe

Tubería hidráulica vertical
(too-beh-REE-ah ee-DRAH-oo-lee-kah vehr-tee-KAHL)

Roof drain

Desagüe de techo
(deh-SAH-gweh deh TEH-choh)

Runoff

Escurrimiento
(ehs-koo-rree-MYEHN-toh)

Self-scouring

Autodesengrasante
(ah-oo-toh-dehs-ehn-grah-SAHN-teh)

Separator

Separador
(seh-pah-rah-DOHR)

Sewage ejector

Eyector de aguas negras
(eh-jehk-TOHR deh AH-gwahs NEH-grahs)

Sewer

Cloaca
(KLOAH-kah)

Alcantarilla
(ahl-cahn-tah-REE-jah)

Sewer, building   **Cloaca de la edificación**
                               (KLOAH-kah deh lah eh-dee-fee-kah-
                               SYOHN)

Sewer, combined   **Cloaca combinada**
                               (KLOAH-kah khom-bee-NAH-dah)

Sewer, public   **Cloaca pública**
                               (KLOAH-kah POO-blee-kah)

Sewer, sanitary   **Cloaca sanitaria**
                               (KLOAH-kah sah-nee-TAH-ryah)

Sewer, storm   **Cloaca pluvial**
                               (KLOAH-kah ploo-VYAHL)

Scupper   **Embornal**
                               (ehm-bohr-NAHL)

Softener   **Ablandador**
                               (ah-blahn-dah-DOHR)

Soil pipe   **Tubo de residuos cloacales**
                               (TOO-boh deh reh-see-DWOHS kloh-ah-
                               KAH-lehs)

**Spillproof
vacuum breaker**   **Interruptor de vacío a
                               prueba de derrame**
                               (een-teh-rroop-TOHR deh)

Stack   **Bajante**
                               (bah-HAHN-teh)

Stack vent   **Respiradero de bajante**
                               (rehs-pee-rah-DEH-roh deh bah-HAHN-teh)

Stack venting   **Ventilación en bajantes**
                               (vehn-tee-lah-SYOHN ehn bah-HAHN-tehs)

Stoppages   **Obstrucciones en tuberías**
                               (ohb-strook-SYOH-nehs ehn too-beh-REE-
                               ahs)

Storm drain   **Desagüe pluvial**
                               (deh-SAH-gweh ploo-VYAHL)

Subsoil drain   **Desagüe del subsuelo**
                               (deh-SAH-gweh soob-SWEH-loh)

**Subsurface
drainage system**   **Sistema de drenaje freático**
                               (sees-TEH-mah deh dreh-NAH-heh freh-
                               AH-tee-koh)

Sump	**Sumidero** *(soo-me-DEH-roh)*
Sump pump	**Bomba de sumidero** *(BOHM-bah deh soo-me-DEH-roh)*
Sump vent	**Respiradero de sumidero** *(rehs-pee-rah-DEH-roh deh soo-me-DEH-roh)*
Supply fitting	**Accesorio de suministro** *(ahk-seh-SOH-ryoh deh soo-mee-NEES-troh)*
Tapping	**Enrosque hembra** *(ehn-ROHS-keh EHM-brah)*
Tempered water	**Agua templada** *(AH-gwah tehm-PLAH-dah)*
Trap	**Trampa hidráulica** *(TRAHM-pah ee-DRAH-oo-lee-kah)*
Trap seal	**Sello de trampa hidráulica** *(SEH-joh deh TRAHM-pah ee-DRAH-oo-lee-kah)*
Tunneling	**Colocación en túneles** *(koh-loh-kah-SYOHN ehn TOO-neh-lehs)*
Vacuum	**Vacío** *(vah-SEE-oh)*
Vacuum breaker	**Interruptor de vacío** *(een-teh-rroop-TOHR deh)*
Vent pipe	**Tubo de ventilación** *(TOO-boh deh vehn-tee-lah-SYOHN)*
Vertical pipe	**Tubería vertical** *(too-beh-REE-ah vehr-tee-KAHL)*
Vent stack	**Respiradero vertical** *(rehs-pee-rah-DEH-roh vehr-tee-KAHL)*
Vent system	**Sistema de ventilación** *(sees-TEH-mah deh vehn-tee-lah-SYOHN)*
Washer	**Arandela** *(ar-ahn-DEH-lah)*
Waste fitting	**Accesorio de desagüe** *(ahk-seh-SOH-ryoh deh deh-SAH-gweh)*
Waste pipe	**Tubo de evacuación** *(TOO-boh deh eh-vah-kwah-SYOHN)*

Waste water	**Aguas residuales** *(AH-gwahs reh-see-DWAH-lehs)*
Water-hammer arrestor	**Disminuidor de golpe de ariete** *(dees-nee-nwee-DOHR deh GOHL-pe deh ah-RYEH-teh)*
Water heater	**Calentador de agua** *(kah-len-tah-DOHR deh AH-gwah)*
Water main	**Tubería maestra de abastecimiento de agua** *(too-beh-REE-ah mah-EHS-trah deh ah-bahs-teh-see-MYEHN-toh deh AH-gwah)*
Water outlet	**Boca de salida** *(boh-kah deh sah-LEE-dah)*
Water pipe	**Tubería hidráulica** *(too-beh-REE-ah ee-DRAH-oo-lee-kah)*
Water service pipe	**Tubería hidráulica de servicio** *(too-beh-REE-ah ee-DRAH-oo-lee-kah deh sehr-VEE-syoh)*
Water supply system	**Abastecimiento de agua** *(ah-bahs-teh-see-MYEHN-toh deh AH-gwah)*
Well	**Pozo** *(POH-soh)*
Well, bored	**Pozo perforado** *(POH-soh pehr-foh-RAH-doh)*
Well, dug	**Pozo excavado** *(POH-soh ehks-kah-VAH-doh)*
Well, drilled	**Pozo taladrado** *(POH-soh tah-lah-DRAH-doh)*
Well, driven	**Pozo hincado** *(POH-soh een-KAH-doh)*
Wet vent	**Tubería húmeda de ventilación** *(too-beh-REE-ah deh OO-meh-dah deh vehn-tee-lah-SYOHN)*
Whirlpool bathtub	**Tina de remolino** *(TEE-nah deh reh-moh-LEE-noh)*
Wrought	**Forjado** *(fohr-HAH-doh)*

121

Yoke vent   **Yunque de ventilación**
(YOON-keh deh vehn-tee-lah-SYOHN)

Tools

Axe	**Hacha** *(AH-chah)*
Ball-peen hammer	**Martillo de bola** *(mar-TEE-joh deh BOH-lah)*
Bar	**Barreta** *(bah-<u>RREH</u>-tah)*
Basin wrench	**Llave pico de ganso** *(JAH-veh PEE-koh deh GAHN-soh)*
Blower	**Sopladora** *(soh-plah-DOH-rah)*
Brace and bit	**Taladro de mano** *(tah-LAH-droh deh MAH-no)*
	Berbiquí y barrena *(ber-bee-KEE ee bah-<u>RREH</u>-nah)*
Broom	**Escoba** *(ehs-KOH-bah)*
Brush	**Pincel** *(peen-SEHL)*
	Brocha *(BROH-chah)*
	Cepillo *(seh-PEE-joh)*
Bucket	**Cubeta** *(koo-BEH-tah)*
	Balde *(BALH-deh)*
Carpenter's apron	**Mandil** *(mahn-DEEL)*
	Delantal *(deh-lahn-TAHL)*
Carpenter's square	**Escuadra** *(ehs-KWAH-drah)*
C-clamp	**Prensa en c** *(PREN-sah en seh)*

Chain pipe wrench	**Llave de cadena** *(JAH-veh deh kah-DEH-nah)*
Chain saw	**Sierra de cadena** *(SYEH-rrah deh kah-DEH-nah)*
Chisel wood	**Escopio** *(ehs-KOH-pee-oh)* **Formón** *(for-MON)* **Cincel** *(seen-SEL)*
Circular saw	**Sierra circular de mano** *(SYEH-rrah seer-koo-LAR deh MAH-noh)*
Circular saw blade	**Disco** *(DEES-koh)*
Claw hammer	**Martillo chivo** *(mahr-TEE-joh CHEE-voh)*
Come-along	**Mordaza tiradora de alambre** *(mohr-DAH-sah tee-rah-DOH-rah deh ah-LAM-breh)*
Combination square	**Escuadra de combinación** *(ehs-KWAH-drah deh kohm-bee-nah-SYOHN)*
Compound mitre saw	**Sierra de corte angular** *(SYEH-rrah deh KOHR-teh ahn-goo-LAHR)* **Sierra de ingletes compuesta** *(SYEH-rrah deh een-GLEH-tehs kohm-PWEHS-tah)*
Cut-off saw	**Sierra para cortar** *(SYEH-rrah PAH-rah kohr-TAHR)* **Sierra de madera** *(SYEH-rrah deh mah-DEH-rah)*
Darby	**Plana** *(PLAH-na)* **Flatacho** *(flah-TAH-choh)*
Drill	**Taladro** *(tah-LAH-droh)*
Drill bit	**Broca** *(BROH-kah)*

	Mecha *(MEH-cha)*
Drill, electric	**Taladro eléctrico** *(tah-LAH-droh eh-LEHK-tree-koh)*
File	**Lima** *(LEE-ma)*
Flashlight	**Linterna** *(leen-TEHR-nah)*
Flat head	**Desarmador de hoja plana** *(des-ar-mah-DOR deh OH-ha PLAH-na)*
Forklift	**Montacargas** *(mohn-tah-CAHR-gahs)*
Framing square	**Escuadra** *(ehs-KWAH-dra)*
Funnel	**Embudo** *(ehm-BOO-doh)*
Goggles	**Lentes de seguridad** *(LEHN-tehs deh seh-goo-ree-DAHD)*
Gloves	**Guantes** *(GWAN-tehs)*
Hammer	**Martillo** *(mar-TEE-joh)*
Hand saw	**Serrucho de mano** *(seh-<u>RR</u>OO-cho deh MAH-noh)*
Hawk	**Esparavel** *(ehs-pah-ra-VEL)*
Hoe	**Azadón** *(ah-sah-DOHN)*
	Zapa *(SAH-pah)*
Hose	**Manguera** *(mahn-GEH-rah)*
Jigsaw	**Sierra de vaivén** *(SYEH-<u>rr</u>ah deh vahy-VEHN)*
Jointer	**Cepillo automático** *(seh-PEE-yoh ah-oo-toh-MAH-tee-koh)*
Jointer plane	**Cepillo de mano** *(seh-PEE-yoh deh MAH-no)*

Knife, utility	**Navaja** *(nah-VAH-hah)* **Cortapluma** *(kohr-tah-PLOO-mah)*
Ladder	**Escalera de mano** *(ehs-kah-LEH-rah deh MAH-no)*
Lawnmower	**Cortadora de césped/pasto** *(kor-tah-DOH-rah deh SEHS-pehd/PAHS-toh)* **Cortacésped** *(kor-tah-SEHS-pehd)*
Level	**Nivel** *(nee-VEL)*
Mallet	**Mazo** *(MAH-soh)*
Mask	**Máscara** *(MAHS-kah-rah)* **Careta** *(kah-REH-tah)*
Medicine cabinet	**Botiquín** *(boh-tee-KEEN)*
Mitre box	**Caja de corte a ángulos** *(KAH-hah deh KOHR-teh ah AHN-goo-lohs)*
Mitre saw	**Sierra de retroceso para ingletes** *(SYEH-rrah deh reh-troh-SEH-soh PAH-rah een-GLEH-tehs)*
Mixer	**Mezcladora** *(mehs-klah-DOH-rah)* **Revolvedora** *(reh-vohl-veh-DOH-rah)*
Nail gun	**Clavadora automática** *(klah-vah-DOH-rah ah-oo-toh-MAH-tee-kah)*
Nail set	**Botador/embutidor de clavos** *(boh-tah-DOHR/ehm-boo-tee-DOHR deh KLAH-vohs)*
Phillips	**Desarmador de punta de cruz** *(des-ar-mah-DOR deh POON-tah deh kroos)*
Pick	**Pico** *(PEE-koh)*

Pick-axe	**Zapapico** *(sah-pah-PEE-koh)*
Plane	**Cepillo** *(seh-PEE-joh)*
Pliers	**Alicates** *(ah-lee-KAH-tehs)*
	Pinzas *(PEEN-sas)*
Pliers, channel lock	**Alicates de extensión** *(ah-lee-KAH-tehs deh ehk-stehn-SYOHN)*
Pliers, vise grips	**Alicates de presión** *(ah-lee-KAH-tehs deh preh-SYOHN)*
	Pinzas perras *(PEEN-sas PEH-rras)*
Plumb bob	**Plomada** *(ploh-MAH-dah)*
Plumb line	**Hilo de plomada** *(EE-loh deh ploh-MAH-dah)*
Pump	**Bomba** *(BOHM-bah)*
Punches	**Punzones** *(poon-SOH-nehs)*
Radial arm saw	**Serrucho guillotina** *(seh-RROO-choh gheeh-joh-TEE-nah)*
Radial saw	**Sierra fija** *(SYEH-rrah FEE-hah)*
Rake	**Rastrillo** *(rahs-TREE-joh)*
Rebar bender	**Doblador de varilla** *(doh-blah-DOR deh vah-REE-jah)*
Reciprocating saw	**Sierra alternativa** *(SYEH-rrah ahl-tehr-nah-TEE-vah)*
Roller	**Aplanadora** *(ah-plah-nah-DOH-rah)*
Router	**Fresadora** *(freh-sah-DOH-rah)*
	Contorneador *(kohn-tohr-neh-ah-DOHR)*

	Buriladora *(boo-ree-lah-DOH-rah)*
Safety glasses	**Gafas de seguridad** *(GAH-fahs deh seh-goo-ree-DAHD)*
Sander	**Lijadora** *(lee-hah-DOH-rah)*
Saw	**Sierra** *(SYEH-rrah)*
	Serrucho *(seh-RROO-choh)*
Saw, electric	**Sierra eléctrica** *(SYEH-rrah eh-LEHK-tree-ka)*
Saw, hack	**Sierra para metales** *(SYEH-rrah PAH-rah meh-TAH-les)*
Saw, hand	**Serrucho de mano** *(seh-RROO-cho deh MAH-noh)*
Saw, power	**Sierra eléctrica** *(SYEH-rrah eh-LEHK-tree-ka)*
Sawhorse	**Burro** *(BOO-rroh)*
Screwdriver	**Destornillador** *(dehs-tor-nee-jah-DOR)*
	Desarmador *(dehs-ar-mah-DOR)*
Sheet metal shears	**Tijeras para metal** *(Tee-HEH-rahs PAH-rah meh-TAHL)*
Shingling hammer	**Martillo para tejamanil** *(mar-TEE-joh PAH-rah teh-hah-mah-NEEL)*
Shovel	**Pala** *(PAH-lah)*
Sledgehammer	**Marro** *(MAH-rroh)*
	Mazo *(MAH-soh)*
Square	**Escuadra** *(ehs-KWAH-drah)*
Solderer	**Soldador** *(sohl-dah-DOHR)*

Soldering torch	**Soplete** *(soh-PLEH-teh)*
Stapler	**Engrapadora** *(ehn-grah-pah-DOH-rah)*
Staple gun	**Engrapadora automática** *(ehn-grah-pah-DOH-rah ah-oo-toh-MAH-tee-kah)*
Strap wrench	**Llave de correa** *(JAH-veh deh koh-RREH-ah)* **Llave de cincho** *(JAH-veh deh SEEN-cho)*
Table saw	**Sierra fija** *(SYEH-rrah FEE-hah)* **Sierra de mesa** *(SYEH-rrah deh MEH-sah)* **Sierra circular de mesa** *(SYEH-rrah seer-koo-LAR deh MEH-sah)*
Thread	**Hilo** *(EE-loh)*
Tool box	**Caja de herramientas** *(KAH-hah deh eh-rrah-MYEN-tahs)*
Trowel, joint filler	**Paleta de relleno** *(pah-LEH-tah deh reh-JEH-noh)*
Trowel, mason's	**Paleta de albañil** *(pah-LEH-tah deh al-bah-NYEEL)*
Trowel, square	**Llana** *(JAH-nah)*
T-square	**Regla T** *(REH-glah TEH)*
Valve seat wrench	**Llave de asientos de válvula** *(JAH-veh deh ah-SYEN-tohs deh VAHL-voo-lah)*
Vice bench	**Torno/tornillo de banco** *(TOR-noh/tohr-NEE-joh deh BAHN-koh)*
Welding mask	**Careta para soldar** *(kah-REH-tah PAH-rah sohl-DAHR)*
Wheel barrow	**Carretilla** *(kah-rreh-TEE-jah)* **Carrucha** *(kah-RROO-chah)*

Engarilla
(ehn-gah-REE-jah)

**Worm drive
circular saw**

**Sierra circular con tornillo
sinfin**
(SYEH-rrah seer-koo-LAR cohn tohr-NEE-joh seen-FEEN)

Wrench

Llave
(JAH-veh)

Wrench, adjustable

Llave francesa
(JAH-veh frahn-SEH-sa)

Wrench, basin

Llave pico de ganso
(JAH-veh PEE-koh deh GAHN-soh)

Wrench, crescent

Llave de tuercas
(JAH-ve deh TWER-kas)

Llave francesa ajustable
(JAH-veh frahn-SEH-sa ah-HOOS-tah-bleh)

Wrench, plumbers

Llave inglesa
(JAH-veh een-GLEH-sa)

Work light

Lámpara de trabajo
(LAHM-pah-rah deh trah-BAH-hoh)

Useful On-the-job Phrases

1. Do you speak English?
   ¿Habla inglés?
   (AH-blah een-GLEHS)

2. What is your name?
   ¿Cómo se llama (usted)?
   [KOH-moh seh JAH-ma (oos-TEHD)]
   ¿Cuál es su nombre?
   (KWAL ehs soo NOHM-breh)

3. My name is.../I am...
   Mi nombre es.../ Me llamo...
   (mee NOHM-breh ehs.../ Meh JAH-mo...)

4. Pleased to meet you.
   Mucho gusto (en conocerlo)
   [MOO-choh GOOS-toh (en koh-noh-SEHR-loh)]

5. What is your phone number?
   ¿Cuál es su número de teléfono?
   (KWAHL ehs soo NOO-meh-roh deh teh-LEH-foh-noh)

6. Please fill out this application.
   Por favor, complete (usted) ésta solicitud.
   [pohr fah-VOR com-PLEH-teh (oos-TEHD) EHS-ta soh-lee-see-TOOD]

7. I need you to fill out this federal tax form.
   Necesito que complete éste formulario de impuestos federales.
   (neh-ceh-SEE-toh keh com-PLEH-teh EHS-teh for-moo-LAH-ree-oh deh eem-PWES-toss feh-deh-RAH-less)

8. And this one for state taxes.
   Y éste de impuestos estatales.
   (EE EHS-teh deh eem-PWES-toss ehs-tah-TAH-less)

9. Also this I-9 form from the government.
   También éste formulario I-9 del gobierno.
   (tam-BYEN EHS-teh for-moo-LAH-ree-oh EE-NWEH-veh dehl go-BYER-noh)

10. I need to see the actual identification you list on the form.
Necesito ver la identificacion que indicó (usted) en el formulario
[neh-ceh-SEE-toh vehr la ee-den-tee-fee-kah-SEEOHN keh een-dee-KOH (oos-TEHD) ehn el for-moo-LAH-ree-oh]

11. Either one from column A or one each from columns B and C.
Se requiere una de la columna A o una de cada una de las columnas B y C.
(seh reh-KYEH-reh oo-nah deh la koh-LOOM-nah AH oh oo-na deh lass koh-LOOM-nas BEH ee CEH)

12. Without I.D., I can't hire you.
Sin la identificación adecuada, no puedo emplearlo.
(seen la ee-den-tee-fee-kah-SEEOHN ah-deh-KWAH-dah noh PWEH-doh ehm-pleh-AR-loh)

13. Do you have a union card?
¿Tiene su credencial de la unión?
[TYEH-neh soo kreh-den-SEE-al deh la oon-YOHN]

14. Can I see it please?
¿Puedo verla por favor?
(PWEH-doh VEHR-la por fah-VOHR)

15. Did the union send you?
Lo manda la unión (el sindicato)?
[loh MAHN-dah la oon-YOHN (ehl seen-dee-KAH-toh)]

16. Can I see the referral.
¿Puedo ver la hoja (referencia)?
[PWEH-doh VEHR-la OH-hah (reh-feh-REHN-syah)]

17. Do you have your own tools?
¿Tiene (usted) sus propias herramientas de mano?
[TYEH-neh (oos-TEHD) soos PROH-pee-as EHR-rah-MYEN-tas deh MAH-noh]

18. If not, I can't use you.
Si no, entonces no puedo emplearlo.
(see NOH, ehn-TOHN-cehs noh PWEH-doh ehm-pleh-AR-loh)

19. Your pay is going to be – per hour.
    Se le va a pagar – por hora.
    (seh leh VAH ah pah-GAHR — pohr OH-rah)

20. ...less tax withholding (...benefits ...union dues).
    ...menos descuentos por impuestos (...beneficios ...cuota de la unión).
    [...MEH-nohs des-KWEN-tos pohr eem-PWES-toss (beh-neh-FEE-syohs ...KWOH-tah deh lah oon-YOHN)]

21. I will pay you at the end (of the day/week-month)
    Le pagaré al final (del día/semana/mes)
    [leh pah-gah-REH ahl fee-NAHL (dehl DEE-ah/seh-MAH-nah/MEHS)]

22. Payday is every Friday (Saturday, Sunday, etc.)
    El día de pago es cada viernes (sábado, domingo, etc.)
    [ehl DEE-ah deh PAH-goh ehs KAH-dah VYEHR-nes (SAH-bah-doh, doh-MEEN-goh)]

23. Can you work tomorrow?
    ¿Puede trabajar mañana?
    (PWEH-deh trah-bah-HAR mah-NYAH-nah)

24. See you tomorrow.
    Nos vemos mañana.
    (nohs VEH-mohs mah-NYAH-nah)

25. Please do not waste materials.
    Por favor no malgaste los materiales.
    (pohr fah-VOHR noh mal-GAHS-teh lohs mah-teh-RYAH-less)

26. Can you drive a car?
    ¿Sabe conducir?
    (SAH-beh kohn-doo-SEER)

27. Do you have a driver's license?
    ¿Tiene licencia de conducir?
    (TYEH-neh lee-SEHN-syah deh kohn-doo-SEER)

28. You may use this bathroom.
    Puede usar este baño.
    (PWEH-deh oo-SAHR EHS-teh BAH-nyo)

29. How late can you work?
    ¿Qué tan tarde puede trabajar?
    (keh tahn TAHR-deh PWEH-de trah-bah-HAR)

30. Are you hungry – thirsty?
    ¿Tiene hambre – sed?
    (TYEH-neh AHM-breh-SEHD)

31. What do you want to eat – drink?
    ¿Qué quiere comer – tomar?
    (keh KYEH-reh KOH-mehr-toh-MAHR)

32. Come with me.
    Venga conmigo.
    (VEHN-gah kohn-MEE-goh)

33. Here is your job safety booklet. Read it and use it.
    Aquí está su folleto de seguridad en el trabajo. Léalo y úselo.
    (ah-KEE ehs-TAH soo foh-YEH-toh deh se-goo-ree-DAD en el trah-BAH-ho. LEH-ah-loh ee OO-seh-loh)

34. Wear these glasses (hat, gloves) for your protection
    Use estos lentes (casco, guantes) para su protección.
    [OO-seh EHS-tohs LEHN-tehs (KAHS-coh, GWAHN-tehs) PAH-rah soo proh-tehk-SYOHN]

35. Are you sick? You need to go home.
    ¿Se siente enfermo? Necesita regresar a casa.
    (seh SYEHN-teh en-FEHR-moh)

36. Are you injured? Go to the doctor/clinic now!
    ¿Se lesionó? ¡Vaya al doctor/a la clínica ahora mismo!
    (seh leh-syoh-NOH. VAH-yah ahl dohk-TOHR/ah lah KLEE-nee-kah ah-OH-rah MEES-moh)

37. Bring the doctor's report when you come back.
Tráigame el reporte (la nota) del doctor cuando regrese.
[TRAHY-gah-meh el reh-POHR-teh (la NOH-tah) dehl dohk-TOHR KWAN-doh re-GREH-seh]

38. Don't use any tools (without asking).
No use ninguna herramienta (sin mi permiso).
[noh OO-seh neen-GOO-nah eh-rrah-MYEN-ta (seen mee per-MEE-soh)]

39. I don't allow beer drinking on the job site.
No se permite tomar cerveza en el área de trabajo.
(noh seh per-MEE-teh toh-mar sehr-VEH-sah ehn ehl AH-reh-ah deh trah-BAH-ho)

40. Drug use is not tolerated.
No se tolera el uso de drogas.
(noh seh toh-LEH-rah ehl OO-soh deh DROH-gahs)

41. Whenever a building inspector shows up, let me know immediately.
Cuando llegue un inspector, avíseme inmediatamente.
(KWAHN-doh JEH-geh oon eens-pec-TOHR, ah-VEE-seh-meh een-meh-DYAH-tah-men-teh)

42. Be careful!
¡Ten(ga) cuidado!
[TEHN(gah) kwee-DAH-doh]

43. Watch out!
¡Cuidado!
(kwee-DAH-doh)
¡Ojo!
(OH-ho)
¡Aguas!
(AH-gwas)
¡Pon(ga) atención!
[POHN(gah) ah-tehn-SYOHN]

44. Follow me.
Sígame.
(SEE-gah-meh)

45. Pull!
    ¡Jale!
    (HA-leh)
    ¡Tire!
    (TEE-reh)

46. Push!
    ¡Empuje!
    (ehm-POO-heh)

47. Bring me that 2x4 (...that fixture, ...that fitting,
    ...that duct, etc)
    Tráigame ese dos por cuatro (...ese accesorio,
    ...esa conexion, ...ese conducto, etc.)
    [TRAHY-gah-meh EH-seh dohs por KWAH-troh (...EH-seh ak-seh SOH-ree-oh, ...EH-sah koh-nehk-SYOHN, etc.)]

48. Help me unload the truck (...lift the beam,
    ...install the drain, etc.)
    Ayúdeme a descargar el camión (...levantar la viga,
    ...instalar el desagüe, etc.)
    [AH-YU-deh-meh ah dehs-car-GAR el kah-MEEOHN (...leh-vahn-TAR la VEE-gah, ...een-stah-LAR el deh-SAH-gweh)]

49. Stack the (lumber/pipe/insulation) over there.
    Apile (la madera/la tubería/el aislamiento) allá
    [ah-PEE-leh (lah mah-DEH-rah/lah too-beh-REE-ah/ehl ah-ees-lah-MYEHN-to) ah-JAH]

50. Cut it at a 45-degree angle.
    Córtelo a un ángulo de cuarenta y cinco
    grados.
    (KOHR-teh-loh ah oon AHN-goo-loh deh kwa-REHN-tay ee SEEN-co GRAH-dos)

51. Hold it there and nail it
    Sosténgalo allí y clávelo.
    (sohs-TEHN-gah-loh ah-GEE ee KLAH-veh-lo)

52. Hold it there while I nail.
    Sosténgalo allí mientras lo clavo.
    (sohs-TEHN-gah-loh ah-GEE MYEHN-trahs loh KLAH-vo)

53. Pick this up.
Levante esto.
(leh-VAHN-teh EHS-toh)

54. Turn it over... Turn it clockwise (counterclockwise)
Voltéelo... Gírelo a la derecha (a la izquierda)
[vohl-TEH-eh-loh... HEE-reh-loh ah la deh-REH-cha (ah la ees-KYEHR-da)]

55. Raise it a little.
Levántelo un poco.
(leh-VAHN-teh-loh oon POH-co)

56. Lower it a little.
Bájelo un poco.
(BAH-heh-loh oon POH-co)

57. That is too heavy. Don't try to lift/carry it alone.
Eso es demasiado pesado. No intente levantarlo/llevarlo solo.
(EH-soh ehs deh-mah-SYAH-doh peh-SAH-do. Noh een-TEHN-teh leh-vahn-TAHR-loh/jeh-VAHR-loh SOH-lo)

58. Get someone to help you.
Pida a alguien que le ayude.
(PEE-dah ah AHL-guee-ehn keh leh ah-YOO-deh)

59. Shovel this into the wheel barrow.
Cargue esto en la carretilla
(KAHR-geh EHS-toh ehn lah kah-rreh-TEE-jah)

61. Put it in the trash bin (or dump truck).
Póngalo en la basura (o camión de volteo).
[POHN-gah-loh ehn lah bah-SOO-rah (oh kah-MYON-deh vohl-TEH-oh)]

62. Clean these... (windows, doors, walls, etc.)
Limpie estas... (ventanas, puertas, paredes, etc.)
[LEEM-pyeh EHS-tahs... (vehn-TAH-nahs, PWER-tahs, pah-REH-dehs, etc.)]

63. Sweep this up.
Barra esto.
(BAH-rrah EHS-toh)

137

64. Hammer this.
Martille esto.
(mahr-TEE-jeh EHS-toh)

65. Where is the saw?
¿Dónde está la sierra?
(DOHN-deh ehs-TAH lah SYEH-rrah)

65. Tie this... (with wire, rope).
Amarre esto... (con alambre, soga)
[ah-MAH-rreh EHS-toh... (cohn ah-LAHM-breh, SOH-gah)]

67. It is break time.
Es hora de descanso.
(ehs OH-rah deh dehs-KAHN-soh)

68. No! Don't do it like that. Please, do it this way.
¡No! Así no lo haga. Por favor, hagalo así.
(NOH. Ah-SEE noh loh AH-gah. Pohr fah-VOHR. AH-gah-loh ah-SEE)

69. Use (the pick, the shovel, the hammer) like this.
Use (el pico, la pala, el martillo) así.
[OO-seh (ehl PEE-koh, lah PAH-lah, ehl mahr-TEE-joh) ah-SEE]

70. Take this to the truck, please.
Lleve esto al camión/camioneta, por favor.
(JEH-veh EHS-toh ahl kah-MYON/kah-myo-NEH-tah, pohr fah-VOHR)

71. Don't drop it, it's very fragil.
No lo deje caer, es muy frágil.
(noh loh DEH-heh kah-EHR, ehs MOOY FRAH-heel)

72. Watch your step.
Cuidado al pisar.
(kwee-DAH-doh ahl pee-SAHR)

The numbers
Los números

0	zero	cero
1	one	uno (una)
2	two	dos
3	three	tres
4	four	cuatro
5	five	cinco
6	six	seis
7	seven	siete
8	eight	ocho
9	nine	nueve
10	ten	diez
11	eleven	once
12	twelve	doce
13	thirteen	trece
14	fourteen	catorce
15	fifteen	quince
16	sixteen	dieciséis
17	seventeen	diecisiete
18	eighteen	dieciocho
19	nineteen	diecinueve
20	twenty	veinte

21	twenty-one	veintiuno
22	twenty-two	veintidós
30	thirty	treinta
31	thirty-one	treinta y uno
40	forty	cuarenta
50	fifty	cincuenta
60	sixty	sesenta
70	seventy	setenta
80	eighty	ochenta
90	ninety	noventa
100	one hundred	cien
101	one hundred and one	ciento uno
200	two hundred	doscientos
300	three hundred	trescientos
400	four hundred	cuatrocientos
500	five hundred	quinientos
600	six hundred	seiscientos
700	seven hundred	setecientos
800	eight hundred	ochocientos
900	nine hundred	novecientos
1,000	one thousand	mil
2,000	two thousand	dos mil
1,000,000	one million	un millón
2,000,000	two million	dos millones

The months of the year
Los meses del año

January	enero
February	febrero
March	marzo
April	abril
May	mayo
June	junio
July	julio
August	agosto
September	septiembre
October	octubre
November	noviembre
December	diciembre

The days of the week
Los días de la semana

Monday	lunes
Tuesday	martes
Wednesday	miércoles
Thursday	jueves
Friday	viernes
Saturday	sábado
Sunday	domingo

UNIT CONVERSION TABLES
SI SYMBOLS AND PREFIXES

BASE UNITS		
Quantity	**Unit**	**Symbol**
Length	Meter	m
Mass	Kilogram	kg
Time	Second	s
Electric current	Ampere	A
Thermodynamic temperature	Kelvin	K
Amount of susbstance	Mole	mol
Luminous intensity	Candela	cd

SI SUPPLEMENTARY UNITS		
Quantity	**Unit**	**Symbol**
Plane angle	Radian	rad
Solid angle	Steradian	sr

SI PREFIXES		
Multiplication Factor	**Prefix**	**Symbol**
$1\ 000\ 000\ 000\ 000\ 000\ 000 = 10^{18}$	exa	E
$1\ 000\ 000\ 000\ 000\ 000 = 10^{15}$	peta	P
$1\ 000\ 000\ 000\ 000 = 10^{12}$	tera	T
$1\ 000\ 000\ 000 = 10^{9}$	giga	G
$1\ 000\ 000 = 10^{6}$	mega	M
$1\ 000 = 10^{3}$	kilo	k
$100 = 10^{2}$	hecto	h
$10 = 10^{1}$	deka	da
$0.1 = 10^{-1}$	deci	d
$0.01 = 10^{-2}$	centi	c
$0.001 = 10^{-3}$	milli	m
$0.000\ 001 = 10^{-6}$	micro	μ
$0.000\ 000\ 001 = 10^{-9}$	nano	n
$0.000\ 000\ 000\ 001 = 10^{-12}$	pico	p
$0.000\ 000\ 000\ 000\ 001 = 10^{-15}$	femto	f
$0.000\ 000\ 000\ 000\ 000\ 001 = 10^{-18}$	atto	a

SI DERIVED UNIT WITH SPECIAL NAMES			
Quantity	**Unit**	**Symbol**	**Formula**
Frequency (of a periodic phenomenon)	hertz	Hz	$1/s$
Force	newton	N	$kg \cdot m/s^2$
Pressure, stress	pascal	Pa	N/m^2
Energy, work, quantity of heat	joule	J	$N \cdot m$
Power, radiant flux	watt	W	J/s
Quantity of electricity, electric charge	coulomb	C	$A \cdot s$
Electric potential, potential difference, electromotive force	volt	V	W/A
Capacitance	farad	F	C/V
Electric resistance	ohm	Ω	V/A
Conductance	siemens	S	A/V
Magnetic flux	weber	Wb	$V \cdot s$
Magnetic flux density	tesla	T	Wb/m^2
Inductance	henry	H	Wb/A
Luminous flux	lumen	lm	$cd \cdot sr$
Illuminance	lux	lx	lm/m^2
Activity (of radionuclides)	becquerel	Bq	$1/s$
Absorbed dose	gray	Gy	J/kg

CONVERSION FACTORS

To convert	to	multiply by
LENGTH		
1 mile (U.S. statute)	km	1.609 344
1 yd	m	0.9144
1 ft	m	0.3048
	mm	304.8
1 in	mm	25.4
AREA		
1 mile^2 (U.S. statute)	km^2	2.589 998
1 acre (U.S. survey)	ha	0.404 6873
	m^2	4046.873
1 yd^2	m^2	0.836 1274
1 ft^2	m^2	0.092 903 04
1 in^2	mm^2	645.16
VOLUME, MODULUS OF SECTION		
1 acre ft	m^3	1233.489
1 yd^3	m^3	0.764 5549
100 board ft	m^3	0.235 9737
1 ft^3	m^3	0.028 316 85
	L(dm^3)	28.3168
1 in^3	mm^3	16 387.06
	mL (cm^3)	16.3871
1 barrel (42 U.S. gallons)	m^3	0.158 9873
(FLUID) CAPACITY		
1 gal (U.S. liquid)*	L**	3.785 412
1 qt (U.S. liquid)	mL	946.3529
1 pt (U.S. liquid)	mL	473.1765
1 fl oz (U.S.)	mL	29.5735
1 gal (U.S. liquid)	m^3	0.003 785 412
*1 gallon (UK) approx. 1.2 gal (U.S.)	**1 liter approx.	
	0.001 cubic meter	
SECOND MOMENT OF AREA		
1 in^4	mm^4	416 231 4
	m^4	416 231 4 10^{-7}
PLANE ANGLE		
1° (degree)	rad	0.017 453 29
	mrad	17.453 29
1' (minute)	urad	290.8882
1" (second)	urad	4.848 137

VELOCITY, SPEED		
1 ft/s	m/s	0.3048
1 mile/h	km/h	1.609 344
	m/s	0.447 04
VOLUME RATE OF FLOW		
1 ft^3/s	m^3/s	0.028 316 85
1 ft^3/min	L/s	0.471 9474
1 gal/min	L/s	0.063 0902
1 gal/min	m^3/min	0.0038
1 gal/h	mL/s	1.051 50
1 million gal/d	L/s	43.8126
1 acre ft/s	m^3/s	1233.49
TEMPERATURE INTERVAL		
1°F	°C or K	0.555 556
		$^5/_9$°C = $^5/_9$K
EQUIVALENT TEMPERATURE ($t_{°C}$ = T_K - 273.15)		
$t_{°F}$	$t_{°C}$	$t_{°F}$ = $^9/_5 t_{°C}$ + 32
MASS		
1 ton (short ***)	metric ton	0.907 185
	kg	907.1847
1 lb	kg	0.453 5924
1 oz	g	28.349 52
***1 long ton (2,240 lb)	kg	1016.047
MASS PER UNIT AREA		
1 lb/ft^2	kg/m^2	4.882 428
1 oz/yd^2	g/m^2	33.905 75
1 oz/ft^2	g/m^2	305.1517
DENSITY (MASS PER UNIT VOLUME)		
1 lb/ft^3	kg/m^3	16.01846
1 lb/yd^3	kg/m^3	0.593 2764
1 ton/yd^3	t/m^3	1.186 553
FORCE		
1 tonf (ton-force)	kN	8.896 44
1 kip (1,000 lbf)	kN	4.448 22
1 lbf (pound-force)	N	4.448 22
MOMENT OF FORCE, TORQUE		
1 lbf·ft	N·m	1.355 818
1 lbf·in	N·m	0.112 9848
1 tonf·ft	kN·m	2.711 64
1 kip·ft	kN·m	1.355 82

FORCE PER UNIT LENGTH		
1 lbf/ft	N/m	14.5939
1 tonf/ft	kN/m	29.1878
1 lbf/in	N/m	175.1268
PRESSURE, STRESS, MODULUS OF ELASTICITY **(FORCE PER UNIT AREA) (1 Pa = 1 N/m^2)**		
1 tonf/in^2	MPa	13.7895
1 tonf/ft^2	kPa	95.7605
1 kip/in^2	MPa	6.894 757
1 lbf/in^2	kPa	6.894 757
1 lbf/ft^2	Pa	47.8803
Atmosphere	kPa	101.3250
1 inch mercury	kPa	3.376 85
1 foot (water column at 32°F)	kPa	2.988 98
WORK, ENERGY, HEAT (1J = 1N·m = 1W·s)		
1 kWh (550 ft·lbf/s)	MJ	3.6
1 Btu (Int. Table)	kJ	1.055 056
	J	1055.056
1 ft·lbf	J	1.355 818
COEFFICIENT OF HEAT TRANSFER		
1 Btu/(ft^2·h·°F)	W/(m^2·K)	5.678 263
THERMAL CONDUCTIVITY		
1 Btu/(ft·h·°F)	W/(m·K)	1.730 735
ILLUMINANCE		
1 lm/ft^2 (footcandle)	lx (lux)	10.763 91
LUMINANCE		
1 cd/ft^2	cd/m^2	10.7639
1 foot lambert	cd/m^2	3.426 259
1 lambert	kcd/m^2	3.183 099